SOUND WAVES

Leonora Davies

SOUND WAVES

Practical ideas for children's music making

CollinsEducational
An imprint of HarperCollinsPublishers

Published 1991 by
CollinsEducational
An imprint of HarperCollins*Publishers*
77–85 Fulham Palace Road
Hammersmith
London W6 8JB

First published in 1985 by Bell & Hyman Limited

Reprinted 1988, 1989, 1991

British Library Cataloguing in Publication Data

Davies, Leonora
    Sound waves: practical ideas for children's
    music making.
    1. School music. Information and study
    1. Title
    372.8′7044   MT9B80

    ISBN 0 00 312529 7

Designed by Geoffrey Wadsley
Illustrated by Mary Budd

Typeset by Inforum Ltd, Portsmouth
Printed in Great Britain by
Scotprint Limited, Musselburgh.

Contents

Acknowledgements

My sincere thanks:
to John Paynter and Murray Schafer whose writings and theories are a constant source of inspiration and are fundamental to my own philosophies;
to John Stephens, to my professional colleagues and particularly to Marjorie Glynne-Jones for stimulating and lively debate which helps to keep ideas and theories refreshed and alive;
to Hilary Matthews, with whom I spent two of the happiest and most formative years of my teaching career;
to friends and colleagues who have contributed more directly to some of the thinking behind the ideas and suggestions in the book: Wendy Bird, Jean Gilbert, Trevor Wishart, Stephen Maw, Margaret O'Shea, Carole Patey, Geoffrey Court and Pauline Adams;
to Linda Gilbert for first teaching me 'The Smuggler's Lullaby';
to my typist Laura Cooke;
to my family for their encouragement and support;
to my editors for their help and guidance
to all the children – many are named in the book – with whom I have been privileged to work; they can teach us so much.

Leonora Davies

The author and publisher would like to thank the following for permission to reprint copyright material: Ted Edwards for the words and melody line of 'Ladybird'; for 'Fog' from *Chicago Poems* by Carl Sandburg, © 1916 by Holt, Rinehart and Winston, Inc, copyright 1944 by Carl Sandburg. Reprinted by permission of Harcourt Brace Jovanovich, Inc; for 'The railway station' by Gwyneth (aged 11) © Oxford University Press 1970. Reprinted from *Wordscapes* edited by Barry Maybury (1970) by permission of Oxford University Press; May Swenson for 'Water picture', © 1956 by May Swenson, originally published in *The New Yorker*, 14 April, 1956, renewed © 1984 by May Swenson.

INTRODUCTION

Learning to listen – both to sounds around them and sounds that they can make themselves – is fundamental to children's music making. By encouraging children to think about these sounds, to be aware of them, to talk about them and to experiment with and manipulate them when they are in groups or by themselves, we provide a framework within which musical activities can take place and a 'real' musical awareness can develop.

This criterion forms the basis for all the practical music-making activities presented in this book. It offers the primary-school teacher and pupil the opportunity to be involved in practical sound and music-making activities. By taking part in this way, I hope that the children will acquire not only the fundamental experiences necessary for musical development but that they will be able to build on these and enjoy participation in musical activities. The classroom teacher with enthusiasm but little 'musical' experience can begin to work confidently with his or her pupils and soon both should benefit. At the same time the music teacher with specialist knowledge is invited to examine his or her current approach to classroom music making. By using the ideas suggested in this book, he or she will be able to create additional opportunities for individual and group music making, so allowing for greater experimentation, decision making and ultimately composition. Above all, the class activities mean that the class can be kept together and music can be made and enjoyed by all the children.

How to use the book

The activities in the book are classified under seven different headings:

Exploratory activities with instruments
Rhythm, pulse and tempo
Vocal activities
Developing instrumental activities
Different starting points
Pentatonic patterns
Shall we sing?

Each section contains *class activities* and *workshop activities*. A great deal of emphasis is placed on the workshop though this should not be attempted until the related class activity has been worked through and understood by the children.

Many of the class activities are structured so that within a corporate situation each child is encouraged to make an individual contribution and decision. However simple this contribution may appear to begin with, it plays a vital role in building up the child's confidence and that is fundamental to a well-balanced general development.

The introduction to each section outlines the principles involved in a particular aspect of music making and suggests any special/specific ways of working that might be necessary.

Each activity is set out in the following way:
* Aim
* Organization and resources
* Activity
* Talking points
* Advice.

A *time guide* is suggested for each activity. The shorter activities (between 5 and 15 minutes) could be used as 'warm-up' activities or as a springboard to others. Many activities will provide the teacher and pupils with material for a complete lesson and developments beyond that.

First names have been used to facilitate the procedural explanations in most of the activities.

It is assumed, however, that the teacher will take the lead in the initial stages of a *class activity* and then invite children to take over.

The activities are graded by stages: 1, 2 and 3. This is an indication of musical experience necessary rather than a specific age group. Most children, however, would not be expected to develop the Stage 3 activities satisfactorily until they are in the third or fourth year of the junior school. On the other hand third or fourth year juniors with no previous musical experience can still enjoy many of the Stage 1 activities. A list of the activities in each stage is given on the next page.

Although all the activities here will provide primary teachers with a great deal of useful material for their practical music making, they will need to supplement this material with other vocal and listening activities in order to achieve a properly balanced primary music curriculum. The teacher will find more useful ideas in several of the books listed in the Bibliography, and of course many radio and television programmes are sources of helpful and stimulating ideas for musical activities in the classroom.

Section	Stage 1	Stage 2	Stage 3
Exploratory Activities with Instruments	How long does my sound last? C.A. How long does my sound last? P.W. Listen to this C.A. Listen to this P.W. What shall we say? C.A. Let's have a chat P.W. How quiet? C.A. How quiet? (game) C.A. Getting quieter C.A.		
Rhythm, Pulse and Tempo	In the gap C.A. Check your pulse C.A. Check your pulse (instrumental) C.A. Names on the click (game) C.A. Patterns and pulses C.A. My turn/your turn C.A.	Names on the click (game) C.A. Names across the circle (game) C.A. Word associations (game) C.A. More patterns and pulses C.A. More patterns and pulses P.W.	
Vocal Activities	Let's all hum C.A. Let's all hum P.W. My turn/your turn (vocal shapes) C.A. ssSSss C.A. Poppp Banggg ShShshsh C.A. Poppp Banggg ShShshsh P.W. Singing names 1 C.A.	Singing names 2 C.A. Singing names 2 P.W. What would you say if . . .? C.A. Pass the hats (game) C.A. Vocal fun C.A. Vocal fun P.W.	
Developing Instrumental Activities		Take your partners please C.A. Take your partners please P.W. Quick whip C.A. What's in my piece? P.W. Da capo P.W. How does it move? P.W.	
Different Starting Points		Contrasts 1 C.A. Contrasts 2 P.W. Reflections P.W.	Reflections P.W. Nobodyes and everybodyes gigge C.A.(2) Nobodyes and everybodyes gigge P.W.(2)
Pentatonic Patterns		My turn/your turn (melodic) Pentatonic patterns P.W. C.A. Pentatonic magic C.A. Pentatonic magic P.W. More magic C.A.	
Shall we Sing?			Using the patterns C.A. Building the patterns P.W. Long John C.A. Long John P.W. Take a chord P.W. Take a tune P.W. Is a song just a song? C.A. Is a song just a song? P.W.

General advice

Organizing instrumental activities with the whole class

* Before the lesson begins instruments and beaters must be laid out ready. A small group of children could be appointed to do this.
* The activity must be conducted in a disciplined and organized way.

These are rules which need to be insisted upon if anything productive is to come out of an instrumental class.

* Children *do not* play when anyone is explaining or discussing the activity. If they fold their arms it might help them to listen more attentively – it will certainly prevent them from fiddling with the instruments.
* Make sure all the children know what the STOP signal is. You may need to practise this.

Operating in a circle is one of the easiest ways of organizing instrumental work in the early stages:

1 Place instruments in the centre ready for use.
2 Children sit in a big circle round the instruments – either on the floor or on chairs if the instruments to be used (such as some pitched instruments) need to be accommodated on tables.
3 When everyone is ready, each child takes the nearest instrument and places it and any necessary beater in front of him/her
4 Arms could then be folded again ready to listen to any necessary instructions.

'All change'
Children welcome the opportunity to play different instruments during the course of one lesson. To facilitate this, *each child passes his instrument and beater to the child on his or her left*. This can be repeated as often as necessary.

Clearing away
Each child puts his instrument and beater down on the floor; three or four children are appointed to tidy up. *Or*: Each child puts his instrument and beater down, then sits quietly with arms folded. Children are called out in groups, one group at a time, to replace their instruments on the trolley (or shelf or cupboard): for example, all those playing skin (or metal or wooden) instruments, or shaking (or banging or scraping) instruments. Beaters are returned to the appropriate pot (felt-headed, rubber, wooden or 'interesting'). The children then return to the circle until everyone is ready.

Teachers are encouraged to devise their own 'rules' for preparing, conducting, and clearing away. These are suggestions only. Organization and discipline, as in a P.E. lesson, are of paramount importance in the smooth running of a class instrumental lesson. Teething problems may occur in the early stages but if the activities are structured and purposeful, if the children have opportunities to develop and experiment with the instruments, then a class lesson need not be a chaotic headache but a rich and productive activity.

Organizing pupil workshop activities

The provision of some kind of 'music area' where children can develop ideas in their own time either individually, in pairs or in small groups, where they can make their own decisions, explore and experiment with sounds, where they are able to develop skills they already have and develop new ones with confidence, is a *vital* extension to their music making. Pupil-initiated work can take place as a class lesson where children work in pairs or small groups, in a large classroom or in the hall. Each group shares its ideas and a general class discussion then takes place.

As a supplement to this way of working or as a follow-up to a teacher-directed activity a small table or area can be set aside in the classroom for children to work in pairs, individually or in small groups on their own during the week.

As well as the structured workshop activities which develop directly from a class activity, children should also be given opportunities to develop their own ideas quite freely. The more workshop opportunities that are provided, the more they will need and want to express their own musical ideas.

When the teacher sets up a workshop as suggested in the book the organization of resources is clearly stated for each activity. These resources should be set out in the 'music area'. The activity can either be explained to the children verbally or it can be prepared in the form of a 'workcard'. The teacher can provide other visual material, pictures or objects which might add further stimulation and encourage other ideas. More ideas about this can be found in the notes which precede the section *Different Starting Points*. Other instrumental resources should be available in case the children find a need to develop the activities in different ways than the original resources suggest.

It is important that children are given the opportunity to share their ideas with the rest of the class, that the teacher monitors the work done in the workshop and that some time for talking about any points the children raise is allowed. This can either be done at the time if flexibility of classroom work allows or a few special minutes set aside every day (at register time?) for 'workshop sharing time'. It is important that teachers are aware of the work being done in the workshop area as it can help in making assessments about each pupil's stage of development and what kind of activity might be most suitable for further development.

Time guide for workshop activities

It is difficult to predict how long children can work productively in the 'music area'. It is important to remember that children need time to work creatively and when they are successfully engrossed in a piece of work they should be given time to 'complete' it. However, in the early stages of working in this way it may be appropriate to allow children 5–10 minutes to explore a few ideas then listen to the ideas 'so far' so that discussion and time for further development are possible. This calls for a flexible approach on the part of the teacher.

Circle activities

The class arrangement for a number of activities is a circle. The following pattern of working can be used for all the activities marked thus: †. This encourages general confidence and musical reassurance by inviting corporate response and then gradually working through to inviting an individual response. It will also help to build up self-confidence in the child who is normally shy and withdrawn.

It may be appropriate initially for the teacher to lead the activities but children should be invited to lead as soon as possible.

1 The leader begins. The others are invited to join in whenever they feel ready. The activity is kept going until everyone, as far as possible, is taking part.
2 The leader begins. The others join in one by one round the circle. At this stage the children learn to anticipate their turn. The leader must always establish which way round the circle the activity will proceed. The activity is completed by children dropping out in turn or round the circle or all finishing together.
3 The leader begins. Individual children are invited to join in by name at random round the circle. In this stage the children cannot anticipate their turn and must be alert and ready to join in if they are called. The teacher must be aware of which children are confident to do this.

Talking points

This is a very important part of the work but can be a little difficult. It can be most effectively done by asking questions about what the children have heard. This method of working helps to encourage both teachers and children to listen actively when others are sharing or performing their compositions. It also encourages the children to think about and question their own as well as others' ideas. An outline for the line of questioning has been suggested for each activity. This is intended as a guide and teachers are urged to develop their own way of discussing the children's work by frequently encouraging them to develop their own ideas. Take care, however, over the way in which this is done, as some children may not be ready for formal discussion in large groups or ready to articulate their experiences along the guidelines suggested.

Use of the tape recorder

Teachers are encouraged to use the tape recorder for children's class work and workshop activities. Children should listen to and comment on their work. Where the workshop activities take place in a 'music area' children can record their own work. A cassette tape recorder could be an integral part of the 'music area'. It also helps teachers to keep an aural record of the children's work and assist in any necessary form of assessment. Take care with the quality of the recording.

Success

As in teaching most subjects the key to success depends largely on:

1 Preparation of material – reading the chosen activity carefully, deciding what will work in your particular circumstances and what may need adapting, and then organizing the resources and arranging the room.
2 The teacher's confidence with the material which will be much greater if the teacher has practice in a new skill or technique before trying it out with the children.
3 The enthusiasm of the teacher – this is perhaps the most important factor.

EXPLORATORY ACTIVITIES WITH INSTRUMENTS

This section is for children with little or no previous experience.

The aim in all these early activities is to familiarize children with the general sound properties and sound qualities of classroom instruments, to stimulate active listening and the sharing of skills and ideas and to encourage a confident enjoyment in instrumental music making.

As soon as children begin to make music in this way they will become actively involved in musical ideas. *Long/short* and *loud/soft* and all the relative contrasts and stages in-between are among the first musical ideas to develop with the children. These should constantly be encouraged and focused on, both in the children's playing and in the discussion about their work. All activities in this section will help to do this.

A time guide is given as a suggestion only for each activity and is indicated thus:

How long does my sound last?

Aim

By playing and listening to the sounds produced on the instruments children can begin to identify and group instruments in a variety of ways.

Organization and resources

Children and teacher sit in a circle on the floor. A variety of instruments – one for each child – placed in front of them in the circle. (See General Advice.)

Activity 1

Everyone picks up their instrument ready to play. They must think how they might make *one* sound on it. Decide which way round the circle the activity will proceed.

* Brenda begins. She makes *one* sound on her tambourine.
* Saleem, the next person round the circle, listens very carefully until the first sound has died away and he can no longer hear it and then immediately makes his sound on his Indian bells.
* The activity continues round the circle with each person making his or her sound in turn in the same way.

Talking points

* 'What did you notice about Mary's sound?'
  'What did you notice about Peter's sound?'
  (The intention is to extract words such as *long* and *short* and *longer than* and *shorter than*.)
* 'Who has an instrument that can make a long sound?'
* 'Can you make a short sound on that instrument?'

Advice

Make sure that the children are listening very carefully all the time.

Each person makes only *one* sound. Each sound must be made *as soon as* the previous sound has died away. The continuity from one sound to another with certain instruments, i.e. woodblock or claves, will need to be anticipated with even more precision than the cymbals or triangles.

Allow the activity to continue round the circle even though some difficulties may be encountered. Discuss any such difficulties afterwards.

'Did anyone notice anything about Sarah's sound? Did she make one or more than one sound? Sarah, try to make just one sound.'

Repeat the activity whilst the children have the same instruments.

Activity 2

'All change.' (See General Advice.) As far as possible each child should have a different instrument. Continue the activity as before.

Activity 3

Proceed as before, but each child must make a sound in a different way on his or her instrument.

Talking points

* 'How was Peter making his sound?'
* 'Can anyone else make a sound like that?'
* 'Was the sound loud (or quiet)?'
* 'Can you make the sound in the same way, but make it loud (or quiet)?'

How long does my sound last?

Aim

Children can record and assess their own findings as they play and listen to sounds made on a variety of instruments.

Organization and resources

Children work in pairs. A variety of instruments that make *long* sounds: finger cymbals, triangles, chimebars, cymbals, gongs. A variety of different beaters: felt, rubber, plastic spoon, toothbrush. Stop watch. Squared paper for block graphs.

Activity 1

Each group chooses two or three instruments and two or three different beaters.
* Taking one instrument and choosing one beater, John strikes the instrument once.
* Peter listens carefully to the length of time he can hear the sound, timing it on the stop watch as he listens. He records the result.
* Repeat this with the same instrument, using the other beaters. Record the results.
* Repeat the whole activity, using the other instruments and the other beaters.
* Record all the findings on a prepared graph. (An example is shown in Figure 1.)
* Let Peter and John talk to the class about their discoveries.

Talking points

* 'How hard did you strike the cymbal?'
* 'Did it make any difference where you struck the cymbal?'
* 'If you strike the same instrument just as hard but in a different place, will the sound last a longer or shorter time?'

Figure 1

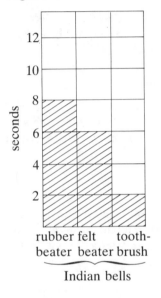
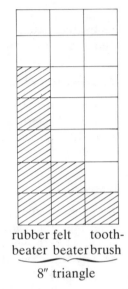

rubber beater   felt beater   toothbrush — Indian bells

rubber beater   felt beater   toothbrush — 8″ triangle

Advice

Encourage different groups to use a different variety of beaters and instruments, so that the information collected and recorded will be varied.

The duration of sound will also depend on how hard the instrument is struck. Allow this discovery to emerge naturally and follow it up through discussion.

Activity 2

Peter and John repeat the activity using the same instruments and beaters, but this time put the instruments on a piece of foam rubber, felt or thick material or put the instruments in a box.

Record the results as before. How does this affect the sounds they hear?

Listen to this

Aim

To provide further activities which help children to explore the sound properties of different instruments.

Organization and resources

Children sit in a circle on the floor. You will need one screen for three or four children to hide behind; a variety of unpitched instruments, one for each child, placed in front of them on the floor.

Activity 1

Choose three or four children (Mary, Camille, Ernest and Pauline).
* They each demonstrate, in turn, a short pattern on their instruments and march off behind the screen, playing as they go.
* When the four children have hidden, the others chant the rhyme:
    Listen to this,
    Listen to this,
    Can you give an answer?
    Loud or quiet, fast or slow,
    Use your ears and you will know.
    Listen to this,
    Listen to this.
* When the children finish chanting the rhyme Mary, behind the screen, plays her woodblock.
* Anyone in the circle who thinks he has an instrument with a similar sound joins in with Mary. He must listen carefully to Mary so he knows when to *stop* playing.
* Chant the rhyme again. This time Pauline plays her chimebar.
* Anyone with a similar sounding instrument joins in.
* Proceed in this way until all four children behind the screen have played.

Activity 2

Proceed as in Activity 1. Beverly, Richard, Tamara and Isabel are behind the screen with their instruments.
* Everyone chants the rhyme.
* Isabel plays a pattern on her tambourine in a particular way.
* Is it loud, quiet, slow or fast? The others must listen first and then respond, echoing the pattern in exactly the same way.
* Tamara then plays a pattern on her cymbal.
* Is it loud, quiet, slow or fast or a combination of these? The others must listen carefully first and anyone in the circle with a similar sounding instrument responds in the same way.

Advice

In order to select the children who will go behind the screen, play a preliminary game.

Choose any simple short song or chant. As the children sing, they pass a drum round the circle. Whoever is holding the drum when the song ends will choose an instrument and be one of the children to hide behind the screen. Repeat this as many times as you need to.

Listen to this

Aim

To provide further opportunity for children to experiment and explore the sound properties of specific groups of instruments.

Organization and resources

Children work in pairs or small groups. One specific group of instruments only should be made available: either instruments made of wood – such as woodblocks of different sizes, xylophones, claves, gato drums; or skin-covered instruments – such as tambourines or drums of various sizes.

Children could be encouraged to collect other sound sources (not conventional instruments) that fit into the particular category in which they are working. Further suggestions for grouping instruments are made in 'Quick whip' in the section *Developing Instrumental Activities*.

Activity

Children experiment on all the instruments. Make up a short piece combining some of the ideas and discoveries they have made. Can they show some of the different sounds that the wooden/skin instruments can make?

Talking points

* 'Did they make loud (or soft) sounds?'
* 'Can you play a pattern which begins softly and gradually gets louder?'
* 'How many different ways did you play the claves?'
* 'Can anyone think of any more ways of making a sound on the claves?'

Advice

Within the structure of the activity – the limitation of the group of instrumental sounds – the children should be allowed to explore and to experiment quite freely. In this way they will demonstrate the skills and abilities they already have and in sharing ideas begin to develop new ones. The discussion that follows work of this kind is crucial in helping children to listen and to think about the sounds they are creating.

What shall we say?

Aim

This activity focuses on the duration of sound – long/short – in connection with the timbre of the instruments. At the same time the activity emphasizes the notions of loud/quiet, through the idea of musical conversations.

Organization and resources

Children sit in a circle on the floor. A variety of unpitched instruments, one for each child, is placed in front of them, on the floor.

Activity

* The children chant the following verse as they pass an 'apple' round the circle:
    What shall we say?
    Whose turn today?
    What shall we say today?
    Are we sad or are we glad?
    Are we bad or are we mad?
    What shall we say today?
* Denise is holding the 'apple' at the end of the rhyme. She has a woodblock in front of her. She plays a pattern on it. It makes a short sound. She must choose a partner who has an instrument that makes long sounds.
* She points to Belinda who has some Indian bells.
* Together they will have a musical conversation. Denise will begin on her woodblock; Belinda replies on the Indian bells.
* The others listen and try to imagine what kind of conversation they are having.

Talking points

* 'What kind of conversation did Denise and Belinda have?'
* 'How did they show this on the instruments?'

Advice

Encourage the children to keep the conversations fairly short for the purposes of the class activity.

You may decide to have a sound signal such as a drum roll, an autoharp chord or a cymbal to signal the end of the 'turn', so the activity can proceed with the rhyme and another pair having a conversation.

Let's have a chat

Aim

To allow children to develop some of the ideas initiated in the class activity 'What shall we say?'

Organization and resources

Children work in pairs. You will need between one and three instruments which make 'short' sounds – such as woodblocks, gato drum, xylophone; and 1–3 instruments which make 'long' sounds – such as Indian bells, triangles, chimebar.

Activity 1

* Saleem and Tun Co each choose an instrument that together will make a contrasting sound.
* Saleem decides on the gato drum. Tun Co prefers the chimebar; he chooses a toothbrush as a beater.
* They decide that Saleem will start.
* Just as in the class activity they have a 'musical conversation'.
* They work on a few ideas and develop these into a short piece.
* They play their 'conversation piece' to the class.

Talking points

* 'How did their conversation begin?'
* 'Did they always take it in turns to speak?'
* 'What kind of mood were the speakers in?'
* 'How did the conversation end?'

Advice

This work can be used to develop ideas both in Drama and in English. Encourage the children to use vocabulary such as 'argument', 'discussion', 'telling off' to describe the kind of conversation.

Activity 2

Work in groups of four.

One pair mimes to the 'conversation' that the other pair creates on the instruments so they are using instrumental sounds to interpret the mood rather than speech.

How quiet?

Aim

To help children become aware of the technical possibilities of instruments by making very quiet sounds on those instruments.

Organization and resources

Children and teacher sit in a circle on the floor. A variety of instruments (one for each person) is placed in front of them on the floor.

Activity 1

Children pick up their instruments. Before playing they think carefully how they might make a very quiet sound.
* They experiment altogether for *one* minute to find a very quiet sound. (What is the STOP signal? See General Advice.) Decide which way round the circle the game will progress.
* Mary begins with her sound. She continues as each person joins in round the circle in turn, playing his own quiet sound, until everyone is playing and listening.
* After a short while Mary stops playing. Everyone else stops in turn, ending with the last player to come in.

Talking points

* 'Are you really making the quietest sound possible?'
* 'Which sounds other than your own could you hear?'
* 'Were any sounds more distinctive than others?'
* 'Why would this be so?'

Advice

Encourage children to listen carefully as they play. Suggest that they listen first to their own sound then try to focus on one other sound in the circle. The self-control needed for the success of this activity must come from within the activity itself.

Activity 2

Repeat the activity but everyone must close their eyes. Children begin to play once they hear that the person next to them is playing. Continue round the circle as before.

Advice

Children may experience some difficulty in playing certain instruments (chimebars or xylophone) whilst they have their eyes closed. Take this into account when preparing the instruments for this activity.

Note: *A pattern is something that we hear or see or feel which is repeated over and over again.*

Activity 3

'All change.' (Whenever possible each child should have a different instrument.)
* Children pick up their instruments. Before playing they should think carefully how they are going to make a very quiet sound.
* Altogether experiment for *one* minute to find a quiet sound.
* Now turn that quiet sound into a pattern. Discuss this with the children (see Note above).
* Allow the children to experiment for *two* minutes. Do they remember the STOP signal? Decide which way round the circle the game will progress.

* Jane begins to play her quiet pattern. She continues and each person joins in, in turn as before.
* After everyone has played together for a short while Jane will stop. Everyone else stops in turn, ending with the last player to join in.

Advice

Before proceeding round the circle, ask one or two children to demonstrate their patterns. This encourages any children who may be unsure of the intentions of the activity. As before, encourage the children to listen to some of the other sounds, as they play.

Some patterns

or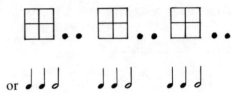

Activity 4

'All change.'
* Ask the children this time to find *two* contrasting ways of making a quiet sound on the instrument they have. Experiment altogether for *two* minutes.
* Turn the *two* quiet sounds into *two* different patterns. Ask one or two children to demonstrate to the others. Decide which way round the circle the game will progress. Misha begins to play his two patterns. The children join in round the circle as before.

Advice

Record the activity. Let the children listen to it immediately afterwards. Can they pick out their own patterns?

How quiet?

Aim

Concentration of listening skills.

Organization and resources

Children and teacher sit in a circle on the floor. A variety of instruments (one for each person) is placed in front of them on the floor.

Activity

* John stands in the centre of the circle.
* Everyone else picks up their instruments.
* Each child in turn demonstrates his very quiet sound pattern.
* John must listen very carefully.
* Blindfold John and turn him round three times.
* Daniel points to Susan and then Peter.
* They play their patterns in turn.
* John must guess who played, which instruments he heard, how the sounds were made, which order they were played in.
* Repeat the game with Ben in the centre and three different children to play their sound patterns.

Advice

The other children must sit and listen very carefully.

Activity 2

Play the game with the children grouped, or sitting or standing in a variety of ways: in a group close together – 'the listener' standing in front of them; spaced out in the hall – 'the listener' standing in different places.

Getting quieter

Aim

A circle activity. Using first body or voice sounds and then instruments, to focus on the idea of sounds gradually getting quieter.

Organization and resources

Children and teacher sit in a circle. Number off in groups of four round the circle.

Activity 1

* Using either body or voice sounds Sally makes a loud pattern.
* Kit repeats the same pattern a little quieter.
* Camilla repeats the pattern even more quietly.
* Ben repeats the same pattern as quietly as possible.
* Mary makes a different loud pattern.
* Philip repeats this pattern a little quieter.
* Proceed round the circle with each No. 1 making a different loud pattern and the subsequent three people repeating this pattern more quietly each time.

Activity 2

* One instrument prearranged in 'sets', in front of each child. Number off round the circle according to the number of instruments in each 'set'; for example, wooden – (1) claves, (2) woodblock, (3) castenets, (4) claves, (5) xylophone, (6) tubular woodblock.
* Sally (1) plays a loud pattern on her claves.
* Each subsequent player *in that group* plays the same pattern but more quietly each time.
* Jane is No. 1 player in the next 'set – of shaking instruments: (1) tambourine, (2) maracas, (3) spray of bells, (4) tambourine.

* The activity proceeds in the same way through this and then the other sets of instruments.

Talking points

* 'Was it easier to get quieter on some instruments rather than on others?'
* 'If this is so, why do you think it is?'
* 'Which group of instruments make the quietest sounds?'

Advice

The instruments can be grouped into 'sets' in a variety of ways. Always encourage the children to arrange this as part of the preparation for the activity. The number of instruments in each group will differ according to your own resources.

RHYTHM, PULSE AND TEMPO

The activities in this section focus on children's rhythmic awareness and provide opportunities for exploring rhythmic possibilities and developing a wider rhythmic vocabulary. They include suggestions for exploratory work based on different pulses and speeds (largely corporate circle activities), and progress to more complex work using combinations of pulses, rhythmic phrases and speeds.

A pulse is a regular beat. It is the underlying feature of a great deal of music that the children are involved in – in their listening, singing and music-making. It can be divided into equal groups of 2, 3, 4, 5, 7 or 9.

Try saying the following sentence out loud: 'Under the wheelbarrow Jennifer sat with a large paper bag placed on top of her head.' Repeat the sentence but accent the syllables that are underlined. The pulse is grouped in '3s', with the accent at the beginning of each group.

Try the same thing with the following proverb:

Here's a little proverb that you surely ought to know,
Horses sweat and men perspire, but ladies only glow.

This time the pulse will be grouped in '4s'.

As an initial introduction to this work with the children, play a steady series of absolutely equal sounds at a moderate speed on a woodblock or claves. Then group these sounds into sets of 4 by accenting the first pulse in each group. You are now playing a repeated pulse pattern of 4. Do the same thing with sets of 3. Try playing the pulses at different speeds. This is the basis of 'Check your pulse' and subsequent 'pulse' activities in this section.

Rhythmic patterns in music vary throughout a piece, usually fitting over an underlying pulse (see 'Patterns and pulses'). Ideas for developing rhythmic vocabulary can be found in 'My turn/your turn'. The emphasis here is on rhythmic phrases rather than isolated rhythm patterns. This makes much more 'musical' sense.

The rhythmic value of beats/notes is traditionally represented in the following way:

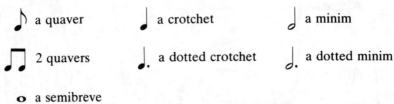

♪ a quaver   ♩ a crotchet   ♩ a minim

♫ 2 quavers   ♩. a dotted crotchet   ♩. a dotted minim

o a semibreve

All the rhythmic examples given in this section assume that the *pulse* is a crotchet. So, '3' means that the 'feel' of the music has a steady repeated crotchet pattern of 1–2–3, 1–2–3, 1–2–3, etc; '5' means that the 'feel' of the music has a steady repeated crotchet pattern of 1–2–3–4–5, 1–2–3–4–5, etc., and so on.

When the pulse beat is a crotchet, then:

♪ = ½ a beat, so ♫ = 1 beat

♩ = 1 beat ♩. = 1½ beats

♩ = 2 beats ♩. = 3 beats

𝅝 = 4 beats

However, the speed of the pulse must *always* be considered. While '4' means that the 'feel' of the music has a steady repeated crotchet pattern of 1–2–3–4, 1–2–3–4, the crotchet pulse will vary according to the indication of the speed – moderately fast; slow; fast (see 'Check your pulse').

This may help teachers to prepare some of the suggested patterns in 'Patterns and pulses' before presenting these activities to the children.

In the gap

Aim

A circle game based on a steady pulse beat. Whilst maintaining a corporate rhythmic structure the game invites individual responses in a variety of ways.

Organization

Children and teacher sit in a circle.

Activity 1

* Fiona sets up a simple pattern – 2 claps followed by 2 gaps, indicated by an open hand gesture – on a steady count of:
  1–2–1–2–1–2–1–2
  x x – – x x – –
* Everyone joins in and keeps it going.
* Decide which way round the circle the game will proceed.
* Fiona begins and says her name into the first gap.
* Jason says his name into the next gap.
* The game proceeds round the circle with everyone in turn putting their names in the gaps and maintaining the clapping.

Advice

Keep the pulse pattern steady and even throughout. Pupils unable to find one gap may be able to find the next one. Although Activity 1 appears to be quite straightforward it is advisable to work through it before going on to the other variations. Say your name in the normal spoken way.

Activity 2

* John sets up the same pattern.
* Everyone joins in as before.

* John says his name into the gap as before but this time he must decide either to 'shout' or 'whisper'.
* Maria says her name into the next gap in the opposite way to John.
* The game proceeds round the circle with everyone alternating with a 'shout' or a 'whisper'.

Activity 3

Try both the activities using different pulse patterns and at different speeds.

Any group of words can be inserted into the gap in the same way (for example, colours, countries, phonic sounds, vegetables). Always set the pulse first before inserting any words. Say the words in the normal spoken way.

(Quite fast) Using phonic sounds:

1	2	3	1	2	3	1	2	3	1	2	3
x	x	x	–	–	–	x	x	x	–	–	–

1	2	3	1	2	3	1	2	3	1	2	3
x	x	x	bell	–	–	x	x	x	bassoon	–	–

(Slowly) Using colours:

1	1	1	1	1	1	1	1	1	1
x	–	x	–	x	blue	x	red	x	burgundy

Useful co-ordinated activities are 'Check your pulse', 'Names on the click' and 'Patterns and pulses'.

Check your pulse†

Aim

An introduction to a variety of pulses and speeds.

Organization and resources

Children and teacher sit in a circle.

Activity 1

* Camilla chooses a pulse pattern and decides
  on a suitable speed, for example:
    patterns of 3 – moderately fast
    or patterns of 4 – fast
    or patterns of 2 – slowly.
* Clap the pattern, accenting the first pulse in
  each pattern, for example:
    1–2–3, 1–2–3, 1–2–3, or
    1–2–3–4, 1–2–3–4, 1–2–3–4.
* The others join in when they are ready.
* Repeat the activity, choosing different groupings and different speeds.
* Work through the two other suggested stages
  for 'Circle activities' given in General Advice.

Advice

Take care to keep the pattern of sounds even
and to maintain the chosen speed (do not slow
down or speed up).

It may help to count the chosen pattern out
loud to begin with. (For further advice refer to
the introduction to this section.)

Activity 2

Introduce a variety of body sounds and movement to the chosen pattern, for example:

1	2	3	1	2	3
clap	click	click	clap	click	click

1	2	3	4
clap	knee slap	click	click

Check your pulse (instrumental)

Aim

An introduction to a variety of pulses and speeds using instruments.

Organization and resources

The children sit in a group, each with an instrument on the floor in front of them.

Activity 1

* Sarah sits in front of the group with a tambour.
* She chooses a pulse group and decides on a speed and begins playing on the tambour.
* The children join in when they are ready.
* Repeat the activity, choosing a different pulse group as before.

Activity 2

* Byron chooses a pulse group and a speed as before and begins playing.
* The children join in when they are ready as before.
* When the pattern is well established Byron changes to a different pulse group.
* Everyone must listen carefully and join in with the new pulse group when they are ready.
* Byron keeps changing. He may repeat some already done.

Activity 3

Repeat the activity with the leader calling on different groups of instruments to play, for example all the tambourines or all the wooden instruments.

The leader must vary the dynamics (volume) and the speed.

Advice

It is important to stress the first pulse in each pattern.

Make sure that most children have picked up one pulse group before changing to a different one.

Names on the click

Aim

A rhythm game to keep them on their toes.

Organization

Children and teacher sit or stand in a circle.

Activity 1

* John sets up a pulse pattern at a moderate speed:

1	2	3	4	1	2	3	4
–	–	–	x	–	–	–	x
clap	clap	clap	finger click				

* Everyone joins in with this and keeps it going throughout the game, taking care to maintain a steady pulse.
* Decide which way round the circle the game will progress.
* Mary begins and says her name on the 'click' of the rhythmic pattern:

	Mary				Peter		
–	–	x	–	–	–	x	
clap	clap	clap	click	clap	clap	clap	click

* Peter must be ready to do the same.
* The game continues round the circle.

Advice

Once the chosen speed is established make every effort to keep it steady and even.

If someone has difficulty in anticipating the 'click' keep the pattern going (within reason) until they can manage to say their name at the right moment.

Activity 2

* Susan sets up the same pulse pattern but introduces some body movements and different body sounds:

		James	
1	2	3	4
knees	knees	knees	knees
slap	slap	slap	bend

		Maria	
1	2	3	4
knees	knees	knees	knees
slap	slap	slap	bend

* Introduce a different speed.
* Alternate the dynamics, sometimes using quiet sounds, sometimes using loud sounds.

Talking points

* 'Did Susan choose a faster or slower speed than John?'
* 'Who can clap a pulse at an even faster (or slower) speed?'
* Ask one or two children to choose and demonstrate pulses at different speeds using different body sounds. Everyone can join in.

Names on the click

Aim

A more complex activity, combining two different pulses.

Organization

As before.

Activity

* John sets up a rhythmic pattern which combines two different pulses, for example:

1	2	3	4	1	2	3
–	–	–	x	–	–	x
clap	clap	clap	click	clap	clap	click

* Decide which way round the circle the game will progress.
* Everyone joins in with the rhythmic pattern.
* Mary begins and says her name on the 'click'.
* They proceed as before.

Further suggestions for combinations of pulses are shown in Figure 2.

Figure 2

1	2	3	1	2		1	2	3	1	2	
–	–	x	–	x		–	–	x	–	x	**3 and 2**
knees slap	knees slap	click	knees slap	click							

1	2	3	4	1	2		1	2	3	4	1	2	
–	–	–	x	–	x		–	–	–	x	–	x	**4 and 2**
shoulder taps			click and knees bend										

1	2	3	4	5	1	2	3	
–	–	–	–	x	–	–	x	**5 and 3**
foot tap				click				

Names across the circle

Aim

More complex circle activities based on 'In the gap' at the beginning of this section.

Organization

Children and teacher sit or stand in a circle.

Activity

* Melissa sets up the clapping pattern. She decides on a clap of 2 and 2 gaps at a fast speed.
* Everyone joins in.
* She says Peter's name into the gap.
* Peter says Brian's name.
* Brian says Susan's name.
* The game continues with names being passed backwards and forwards across the circle until someone misses a gap. The 'culprit' must then do a musical forfeit.
* The game starts again.
* This activity can also be done using the idea of alternating the dynamics (variations in loud/ quiet).

Suggestions for musical forfeits
* Choose a song and sing a verse very slowly or very quickly.
* Sing through 'Twinkle, twinkle, little star', missing out the last word of each line.
* Sing your name pattern, going up in scale from as low to as high as you can.

Word associations

Aim and Organization

As in 'Names across the circle'.

Activity

* Tom sets up a clapping pattern. He decides on
  a clap of 3 and 3 gaps at a slower speed than
  Melissa.
* Everyone joins in.
* Tom thinks of a word. He says it into the gap.
* Proceeding round the circle Mary must say into
  the next gap a word which has some association
  with Tom's word.
* Jane does the same. (See Figure 3.)

Figure 3

1 2 3	1 2 3	1 2 3	1 2 3
x x x	– – –	x x x	– – –
(clap)	banana		monkey

1 2 3	1 2 3	1 2 3	1 2 3
x x x	– – –	x x x	– – –
	zoo		animals

Patterns and pulses†

Aim

To stimulate and develop rhythmic vocabulary.

Organization

Children and teacher sit in a circle.

Activity 1

* Francis claps a repeated rhythmic pattern, for example:

* The others join in when they are ready.
* When the pattern is well established, Francis calls 'All change!' and without stopping begins a different pattern:

* The others pick up the new pattern as soon as they can.
* Francis continues in this way for two or three minutes, while calling 'All change!' and introducing different patterns.

Figure 4

* Once the idea is established introduce a variety of hand and body sounds.
* Repeat the activity and establish different pulse speeds (faster or slower) than the previous example.
* Repeat the activity with different leaders once the children are experienced and confident.

Advice

A pulse group of four may be the easiest to work with initially.

The teacher will need to lead this activity until the children are more experienced.

Activity 2

* Introduce rhythmic patterns using pulses of 3, 5, 6 and 7 in the same way.
* Using just one repeated rhythmic pattern, work through the other two suggested stages for 'Circle activities' (see General Advice).

Different rhythm patterns

Some suggestions are given in Figure 4 for more rhythm patterns using other pulses. Remember to vary the speed, but demonstrate and establish the new speed each time.

More patterns and pulses†

Aim

Further work using different pulses and rhythmic patterns and aurally combining the two.

Organization

Children and teacher sit in a circle.

Activity

* As a class decide on the pulse. Will it be 2, 3, 4, 5, 6, or 7?
* Decide on a speed and clap the pulse. Let everyone join in with this for a few seconds.
* Using a variety of body sounds choose a small group who will keep this going, for example:

3 ♩ ♩ ♩ ‖
   clap click click

* The 'pulse' group begins.
* James claps a rhythm pattern that will fit with the chosen pulse:

* The others join in when they are ready.
* When the pattern is well established, James calls 'All change' and without stopping begins a different pattern.
* The 'pulse' group continues.
* The others pick up the new pattern as soon as they can.
* Continue the activity as before.
* Using just one repeated rhythmic pattern, work through the other two suggested stages for 'Circle activities' (see General Advice).

Advice

Use a variety of different pulses.

Let everyone clap the new pulse each time before dividing into the two groups.

More patterns and pulses

Aim

To provide an opportunity for children to use instruments and make up a short piece demonstrating the use of two different pulses.

Organization

Children work in pairs or small groups. Each child has an instrument.

Activity

* The group decides on a pulse.
* Camilla plays the pulse on a woodblock.

3 ♩ ♩ ♩ | ♩ ♩ ♩ ‖

* Jane makes up a rhythm pattern that fits with the pulse. She plays it on the drum.

3 ♫ ♩ ♩ | ♫ ♩ ♩ ‖

* Susan joins in with the same rhythm pattern on the chimebar.

3 ♫ ♩ ♩ | ♫ ♩ ♩ ‖

* The girls choose a different pulse and experiment in exactly the same way.

4 ♩ ♩ ♩ ♩ | ♩ ♩ ♩ ♩ ‖

4 ♫ ♩ ♩. ♪ | ♫ ♩ ♩. ♪ |

* Now they can use their ideas to make up a short piece using the two different pulses.
* Let the groups perform their pieces to each other.

Advice

Make sure the children know how they are going to change from one pulse to the other.

Decide how many times Jane and Susan will play their first pattern before Camilla changes to a different pulse. The speed that they choose may be quite an important factor.

Talking points

* 'Which two pulses did the girls choose?'
* 'How did they change from one pulse to the other?'
* 'Did anyone use their instruments in an interesting (or particular) way?'
* 'Did they have a variety of loud and soft playing?'

◐ Activity

◑ Performance and discussion

My turn/your turn (rhythm)

Aim

Further opportunities for encouraging and developing rhythmic phrases.

Organization

Children sit comfortably in a group.

Activity 1

* Samantha chooses a speed and demonstrates by clapping the pulse beat at that speed to the class.
* She chooses a pulse of 4 at a moderate speed.

* She then claps a rhythmic phrase based on that speed and pulse which the children immediately echo back.
* Each rhythmic phrase fits over 2 pulse patterns (see Figure 5).
* Samantha continues with different phrases for two or three minutes without stopping.
* Repeat the activity, inviting different pupils to take the lead.
* Introduce a variety of hand and body sounds and different dynamic (loud/soft) possibilities. Try the same phrase first loud, then soft, or a phrase that begins softly and gets louder.

Advice

Begin with simple phrases, but introduce more difficult ones according to the experience of your children.

Make sure the children are listening carefully and responding accurately.

A steady pulse of 4 may be the easiest to work with initially, but introduce rhythmic phrases using pulses of 3, 5, 6 and 7 in the same way. See 'Patterns and pulses' for rhythmic suggestions.

Activity 2

* Valerie chooses three or four different hand and body sounds.
* She demonstrates these to the class, for example:
    hand clap – loud
    hand tap – soft
    knee slap
    stroking palms together (see Figure 6).
* She demonstrates the chosen pulse and speed.
* The other children now close their eyes.
* Valerie proceeds as before.

Advice

The leader, whose eyes are open, must look at the others carefully. Some children may not be able to echo back the phrases quite accurately to begin with.

Try and repeat any phrases that they find difficult.

Can they hear the difference between a hand slap and a knee slap?

See 'Patterns and pulses' for other rhythmic ideas.

Figure 5

Figure 6

VOCAL ACTIVITIES

Singing is a fundamental part of children's music making. The activities in this section, however, encourage teachers to take a broader view of the range of vocal activities possible. Some of the activities focus on using the voice in an imaginative and less formal way; other activities involve work using specific notes to encourage and develop more precise pitch awareness. Many of the activities give the teacher an opportunity, in the classroom situation, to assess individual children's sense of pitch. Younger or less experienced children may have a smaller vocal range than older and more musically experienced children.

Be aware of any child who seems to have a particular problem in singing or pitching notes. The reasons for this could be many and complex:

Is the note outside the child's vocal range?
Is he or she listening carefully? Is there a hearing problem?
Is he or she lacking in confidence?

Constant encouragement within the classroom situation will go a long way to helping children develop confidently in vocal work.

Children begin to record some of their ideas using picture shapes, word or phonic patterns (see 'Let's all hum' and 'ssSSss'). If teachers wish, they could begin to present the conventional notation to their pupils.

Notation is a visual representation of the sounds created; 'picture' shapes and conventional crotchets and quavers are both ways of writing sounds down. In these early stages of music making, it is essential that any visual representation is made *as a result of* or *in conjunction with* the actual sound.

Pitch is traditionally represented in the way shown in Figure 7.

Figure 7

Middle C

C₁ D₁ E₁ F₁ G₁ A₁ B₁ C D E F G A B C'

All the notated examples used in this book can be 'worked out' by using this stave as a reference point. Teachers for whom this is an exciting new departure are encouraged to play these notes on the piano or alto xylophone.

For other musical signs and terms please refer to the Appendix.

Let's all hum†

Aim

A useful 'warm-up' activity for the beginning of a lesson which helps to promote a relaxed and co-operative atmosphere.

Organization

Teacher and children sit in a circle.

Activity 1

* Samantha thinks of a note.
* She hums that note and sustains it.
* When they are ready everyone joins in on the same note.
* Samantha changes to a different note.
* The others join in with the new pitch.
* Samantha continues with the activity, pitching three or four notes.
* Invite a different child to lead.
* Repeat the activity, working through stages 2 and 3 of 'Circle activities' (see General Advice).

Advice

Try the activity with closed eyes: (a) children may be less inhibited; (b) it helps to focus their listening attentions and concentration.

The leader must listen carefully to the others in order to judge when to change to a new pitch.

Not all the children will pitch the notes accurately at first. The activity gives the teacher an opportunity to assess her or his pupils' sense of pitch development.

Talking points

* 'Did Samantha begin on a low, or a high or a middle range note?'
* 'Who can trace in the air or on the board the way in which the notes moved from one to the other?'

Activity 2

Repeat Activity 1 with Tracy as leader, who chooses three or four notes. Encourage the children to vary the pitch dramatically.
* Can Peter draw the shape on the board as she hums?
* Using the picture shape on the board, can Jenny hum the notes that Tracy chose?
* Build up other 'graphic pictures' in the same way, for example:

Let's all hum

Aim

An opportunity for children to re-create the graphic pictures they built up in the class activity.

Organization and resources

Children work in pairs. Children's picture scores available.

Activity

* Ben and Katie choose David's picture score:

* Ben decides to sing the notes to an 'ooo' sound.
* Katie points to the notes.
* Ben sings each note as she points.
* Katie decides with her pointing how long Ben will hold each note on for.
* They change over.
* Katie decides to sing her notes to an 'eee' sound.
* They choose another picture score to work on.

My turn/your turn (vocal shapes)

Aim

To encourage imaginative and experimental vocal sounds.

Organization

Children sit in a group. Samantha stands in front.

Activity 1

* Samantha thinks of an interesting sound that she can make with her voice. She makes it, for example:

$$Bah \longleftarrow Tттт$$

* The children listen carefully and repeat the sound exactly.
* Samantha makes a different sound.
* The children copy the sound.
* Samantha continues in this way, making a variety of different sounds with her voice which the others copy (see examples in Figure 8).

Figure 8

triiingg eeeeeeeeee
bbbb — bbb   POW

* Repeat the activity with different leaders.
* Try the activity with closed eyes.

Talking points

* 'Who can make a sound using their mouths but not their voices?' – This may encourage sounds with tongue, teeth and lips.
* 'Who can make sounds which begin quietly and get louder?'
* 'Who can make some very quiet sounds and one very loud sound?'

Advice

By questioning the children in this way the teacher is also indirectly making some suggestions and giving the children ideas for developing the activity.

Choose the leaders with care. Some children may be a little diffident at first.

ssSSss

Aim

To experiment and explore imaginative ways of using voices, with a particular emphasis on getting louder and softer.

Organization and resources

Children sit in a group, later to divide into two or three groups. Maria the conductor stands in front of the group. Paper, felt pens or blackboard and chalks will be needed.

Activity 1

* Choose a phonic sound, for example SSSSSS.
* Using hand signs, Maria conducts the class (see 'Useful conducting signs' in the section *Developing Instrumental Activities*).
* She begins quietly, gets louder then quieter again, for example:

sssSSSSSsssss

* Some children can draw the shape of the sound as it is made.
* Choose other sounds, for example MMMMMM or OOOOOO. Maria conducts the class as before. Children draw these sounds in the same way (see examples in Figure 9).

Advice

The children must watch the conductor very carefully.

Encourage the conductor to experiment with the speed of the dynamic changes, for example:

Start loud and get quieter very quickly.

Start quietly and very gradually get louder.

Encourage the children to use *silences* – start loud, using the appropriate hand sign, stop the class then start again quietly. This will also make the 'picture shapes' look more interesting.

Activity 2

This develops the vocal ideas started in Activity 1.

* Divide the class into two or three groups. Each group chooses a different vocal sound and follows its own conductor.
* Using the appropriate hand signs the conductor controls how and when his group will make their sound.
* Encourage the ideas of dynamics and speed as in Activity 1.

Figure 9

M M M Mmmmmmmm

or mmmmm M M M M

O ooo   oooo

PPP P P   PP P P   PPPP   P

Advice

Encourage the children to listen to the other sounds as they are making their own.

It may be appropriate to repeat the activity once or twice with the same sounds and the same conductors before changing round.

Activity 3

Each group chooses two phonic sounds. The conductor will point to a prepared card to indicate which sound the group will make (see Figure 10 for examples).

Proceed in the same way as before. Record the activity and let the children listen to the results.

Figure 10

rrrr rrr r r

BBB bbb b B

Cr Cr Cr cr cr cr Cr

Talking points

* 'How did Peter's group start their sound?'
* 'Which group made very loud sounds?'
* 'What happened in the middle of the sound Saleem's group were making?'

The nature of these questions will depend upon the way in which they choose to make their sounds.

Poppp Banggg ShShshsh

Aim

Further experiments with sound words and picture shapes, focusing on the musical ideas of loud and soft.

Organization and resources

Prepare a few visual examples from comic strips to give the children initial stimulus (see Figure 11).

Paper, felt pens, paints or crayons will be needed.

Figure 11

Activity

* Discuss the visual examples with the class.
* Make some of the sounds.
* Suggest grouping the sounds, for example, loud/quiet.
* Encourage imaginative ways of vocalizing the sounds.
* Ask the children to suggest a few examples.
* Let them demonstrate their sound to the class, illustrate their suggestions and then conduct the rest of the class making their sound.
* Collect one or two more suggestions in this way.

Organization

Children divide into small groups with paper and felt pens etc.

Activity

Let them discuss and share some more sound words and draw the shapes.
* Using four or five of their sound words each group will prepare a short vocal piece.

* They may like to choose only quiet sounds or only loud sounds.
* How will they know when to make their sounds?
* Do they need a conductor or will they watch and listen to each other carefully?
* Let each group perform their piece.

Singing names 1†

Aim

To encourage children to listen and then to sing patterns based on two notes of definite pitch.

Organization and resources

Everyone sits in a circle. Saleem sits in the centre ready to play notes E and G on chimebars, alto xylophone or alto glockenspiel. These instruments present a pitch which relates to the children's voices.

Activity 1

Saleem decides on a speed and plays the two notes as a steady pulse at that speed, for example:

* Everyone listens carefully.
* When they are ready they hum the two notes as Saleem continues to play them.
* When Saleem stops playing everyone stops humming.
* Saleem begins again. This time he may choose to play the two notes the other way round, i.e.:

* The children listen carefully and join in humming as before.
* Let Saleem have two or three more turns.
* How can he vary the patterns this time? Use only steady beats in patterns (see Figure 12).

Figure 12

* Children close their eyes. Repeat the activity.
* Repeat with different leaders.
* Repeat the activity, working through stages 2 and 3 of 'Circle activities' in General Advice.

Talking points

* 'Which way did Saleem's first pattern move?'
* 'Can you trace the shape of it in the air?'
* 'Can you hum the pattern *without* the chimebars?'

Advice

The patterns given as examples are based on a 2- and a 4-pulse group. Be prepared for children to produce more imaginative patterns, perhaps based on a 3- or a 5-pulse group.

Children must listen very carefully and stop when the leader stops. They must remain still, quietly anticipating until they hear the new pattern.

Activity 2

* Jeremy decides on a speed and plays a melodic pattern using the two notes, at a steady pulse beat, for example:

This pattern is based on a pulse of 3.

* Everyone listens very carefully.
* When they are ready they sing their name patterns using the pitch pattern of the two different notes that Jeremy is playing but making up their *own* rhythmic pattern according to their names.
* When Jeremy stops everyone stops singing.
* He begins with a different pitch pattern and the class joins in when thcy arc ready.
* Proceed as before.
* Children close their eyes. Repeat the activity.
* Repeat with different leaders.
* Work through stages 2 and 3 of 'Circle activities' in General Advice.

Figure 13

(a)

(b)

Advice

Some children will have names that fit easily over the steady pulse pattern (Mary, Susan, Sarah); others with more complex patterns (Penelope, Rebecca, Leonora, Mahendrath) may naturally find ways of fitting their name patterns over the pulse.

If they have any difficulty encourage a class discussion. Ask another child how he might sing the name. You may have to shorten the name (with the child's permission) for the purposes of the activity.

Children with only one syllable to their name may decide to slur their pattern between two notes (see Figure 13a) or they may repeat the name pattern over the number of notes in the pulse pattern (as shown in Figure 13b).

Singing names 2

Aim

Further development of the previous activity. To sing name patterns based on two notes against a melodic ostinato using three notes. (Note: An ostinato is a rhythmic or melodic pattern of notes, which can consist of one note or up to three or four notes, which is repeated throughout.)

Organization and resources

Everyone sits in a circle. Benita sits in the centre ready to play notes C, D and E on an alto xylophone. Mary sits in the centre ready to play notes E and G as before on chimebars.

Activity

* Benita makes up an ostinato pattern, using the notes C, D and E. She keeps a steady pulse of four beats, for example:

* As she continues to play this the class joins in and sings the pattern to 'la'.
* When she stops everyone stops singing.
* Mary plays a pattern using the two notes E and G. Use only steady beats in the pattern, for example:

* As she plays the others 'think' their name patterns in their heads.
* Benita begins. She plays her ostinato pattern at a steady pulse.
* Mary joins in with her pattern. She must listen carefully and join in with her pattern on the first beat of the ostinato pattern.
* The others listen carefully to Mary and sing their name patterns as before, using the pitch direction that Mary indicates (see Figure 14).
* When Mary stops the children stop.
* Benita continues her ostinato pattern.
* Mary begins again telling the class with her chimebar pattern how they will sing.
* The activity continues as before.
* Work through stages 2 and 3 of 'Circle activities' given in General Advice.

Advice

The introduction of a second melodic phrase makes this activity quite sophisticated for some children.

It is important to have the two patterns played on instruments with different timbres. It helps the children to identify them. If bass xylophone bars are available these would be even better for the ostinato.

Figure 14

Singing names 2

Aim

To provide an opportunity for the children to develop and to use the ideas covered in the class activity and to make up a short piece using some of their ideas.

Organization

Children work in small groups, ideally of four children or more.

Make available the notes C, D and E on one instrument and the notes E and G on another.

Activity

* Mary makes up and plays the ostinato using the notes C, D and E. This could be extended to notes G and A making up a pentatonic scale for those children who are musically experienced (see section *Pentatonic Patterns*).
* Peter plays the notes E and G.
* Susan and Saleem sing their name patterns as before. Encourage them to discuss and share their own ideas.
* Working on their own ideas and some from the class lesson, they make up a short piece using their name patterns and an ostinato.
* Let the groups perform their pieces to each other.

Talking points

* 'Did Susan and Saleem sing their name patterns at the same time or did they take turns?'
* 'Were there any contrasts between playing and singing loudly and softly?'
* 'How fast (or slow) was the piece?'

Advice

Allow the children to organize themselves within the group, i.e. who plays what. They may like to change round after a while.

After the discussion time groups may wish to work further on their pieces.

◗ Workshop

◗ Discussion

◗ Improvement time

What would you say if . . . ?

Aim

To encourage children to use their singing voices in a free and uninhibited way.

Organization and resources

Children sit or stand spaced out in the hall or a large classroom.

Activity

Ask the children to imagine that they are feeling angry about something. What kinds of things might they say?

* At a given signal (a cymbal, perhaps) everyone sings quite freely what they are feeling.
* Let the activity continue for a very short time.
* Choose other moods or situations which might encourage different ways of singing – loud/soft, fast/slow.
* The children could feel tired, anxious, excited or sad.
* Once the children gain confidence they could develop their ideas in pairs and have a 'sung conversation or argument'.

Advice

Encourage a class discussion about the kinds of things one might say in these various situations. This may help those children who are less confident.

Develop some of these ideas in English and Drama lessons.

Pass the hats or
Question and answer

Aim

An opportunity for children to combine vocal and dramatic ideas in a singing conversation, using notes of a defined pitch.

Organization and resources

Children and teacher sit in a circle on the floor. Two hats – these must indicate certain characters, for example a policeman and an old lady – are needed.

Notes E and G on a xylophone. Notes E and G on chimebars ready in the centre of the circle.

Tun Co sits with his back to the class with a drum.

Activity

* As Tun Co plays the drum the two hats are passed round the circle in opposite directions.
* When the drum stops Sally has one hat on and Philip has the other.
* They go to the centre of the circle: Philip will ask the questions; Sally will reply.
* The two children play the two notes as a steady pulse beat. The questions and answers will be sung quite freely over this, using the pitches of the two notes.
    'What can I do for you, Madam?'
    'Well, you see, I've lost my cat.'
    'What colour is it?'
    'It's a black one with white tips on his paws.'
* Tun Co listens carefully to the conversation and after a short time begins to play the drum again.
* Philip and Sally return to the circle and the game proceeds.

Advice

The main intention of the game is to encourage a sung dialogue. If some children have difficulty in pitching their singing round the two notes, do not let this inhibit the dialogue. Encourage them to listen to the notes they are playing. Encourage the children to use complete sentences.

Vocal fun

Aim

To combine the ideas suggested in the experimental vocal work and those of pulse to create a class piece.

Organization

Children sit in a group, later to divide into four groups. James the conductor stands in front of them.

Sets of cards prepared with different phonics using a variety of pulses (Figure 15) are needed; also four music stands.

Activity

* As a class decide on a pulse. Select the appropriate set of cards.
* Vanessa plays the chosen pulse on a woodblock, accenting the first beat in each group.
* Children tap the pulse quietly.
* Vanessa continues to play throughout.
* James holds up one card and conducts the class in on the first beat of the pulse. The class make the appropriate phonic sounds, keeping in time with the pulse.
* Repeat this. James will tell the children when to stop and start by using the appropriate hand signs.
* Repeat this with the other three cards. It is important that all the children should practise all the patterns.
* Divide the class into four groups.
* James puts one card on a music stand in front of each group.
* Vanessa continues to keep the pulse going quietly on the woodblock.
* James conducts the first group in. They continue with their pattern.
* He brings in the other groups in turn in the same way.

* Each group must watch James as he will tell them when to stop.

Talking points

* 'Can we think of different ways of finishing our piece?'
* 'Shall we all finish together or one group at a time?'
* 'Is this pattern more effective if we say it loudly (or quietly/getting louder/getting quieter)?'
* 'Which group shall we begin with?'

Advice

It may be easier for each group to come in in turn and finish in the same way, but encourage the children to suggest different ways of shaping the piece. Always be ready to encourage dynamic qualities.

Use the other pulse patterns in the same way.

Vocal fun

Aim

To allow children the opportunity to reinforce and to develop the ideas used in the class activity.

Organization

Children in groups of four or five. One instrument suitable for keeping the pulse. Paper and felt pens.

Activity

* The group decides on a pulse.
* Mary plays this on a suitable instrument.
* Peter, Toby and Robyn experiment in turn with various phonic and vowel patterns that fit over the chosen pulse.
* Each one decides on a pattern that they are happy with and writes the pattern clearly on a piece of paper.
* Mary plays the pulse.
* The others experiment with their chosen patterns.
* They make up a group piece using some of the ideas from the class activity and class discussion. How will the piece end?
* Let each group perform their piece.
* Display the visual patterns on the classroom wall.

Advice

When working on their own some children may have difficulty in coming in on the first beat of each pulse pattern. Encourage them to listen carefully to the instrument playing the pulse pattern and to try counting with the pulse before they come in with their pattern.

Figure 15
Some examples of rhythmic phonic patterns using different pulse bases

3	um	cha	cha
3	zoom	⟶	pah
3	ᴦᴦ	ᴦᴦ	K↗
3	clink	—	—

4	ch	s s	mah
4	b̲b̲	toot ⟶	b
4	do —ng	do —ng	
4	um — chacha	K↗ pah	

5	ch	ch	ᴦᴦ	K P
5	b̲b̲	b̲b̲	toot↗	b b
5	s	s	s̲s̲	um↗ pah
5	pah ⟶			b̲b̲

Prepare four separate cards for each rhythmic set.

DEVELOPING INSTRUMENTAL ACTIVITIES

This section is for children with some previous musical experience.

These activities continue to offer children the opportunity to explore sounds and to experiment with the instruments but they are now encouraged to develop, to use and to structure their ideas into musical patterns, phrases and combinations of patterns. Many of the activities suggest ways of helping children to shape these explorations into compositions.

Children now begin to explore musical ideas involving speed and pitch and all the relative contrasts and stages in-between. This section includes examples of workcards for 'music area' activities. A model of wording and layout is suggested. Picture representations are used and developed as a way of helping the teacher to provide both structured and informative activities for the children to develop.

Take your partners please

Aim

To sort and group sounds of similar and contrasting timbres.

Organization

In the hall or a large classroom with space to move.

A variety of hand-held percussion instruments, one for each child, arranged round in a circle. The choice of instruments is *important*. Make sure there are 'pairs' of instruments or instruments with similar sounds, i.e. woodblock/claves or triangle/finger cymbals. One tuned instrument – recorder or xylophone; *or* record player/piano/tape recorder.

Activity 1

* Martin plays or improvises a tune on the recorder or xylophone (*or* the teacher plays a piece on the piano/record player etc.).
* Children move round the instruments in response to the music. The music could indicate to the children how to move, i.e. skip, run, walk, fast, slow.
* When the music stops each child picks up the nearest instrument.
* Standing still everyone makes up a short pattern on the instrument they are holding.
* At a given signal (a cymbal bang from Mary) each child, playing his instrument, moves towards someone playing a similar-sounding instrument.
* Once in pairs the children make up a short piece together, combining the patterns they have made up.

Activity 2

* Listen to each pair.
* Replace the instruments in a circle.
* Repeat Activity 1.

Talking points

* 'What kind of sounds can you hear?'
* 'How did Robert play his instrument?'
* 'Did you notice anything special about Peter and Tom's piece?'
* 'How did Sara and Melanie's piece end?'

Advice

Let a different child play the 'moving round' tune. Hopefully the children will have an opportunity to play a different instrument.

Activity 3

* John plays or improvises a tune on the recorder or xylophone (*or* the teacher plays a piece on the piano/record player, etc.).
* Children move as before in response to the music, round the instruments.
* When the music stops each child picks up the nearest instrument.
* Standing still, everyone makes up a short pattern on the instrument they are holding.
* At the given signal each child, playing his instrument, moves towards someone playing an instrument making a contrasting sound.
* Once in pairs the children have a musical conversation, i.e. first one plays, then the other.
* Listen to each pair.

Talking points

* 'What kind of conversation were Jane and Susan having?'
* 'Could we tell what sort of mood they were in?'
* 'How could we tell that Martin was angry?'
* 'Were there any silences in their conversation or did they talk all the time?'

Advice

Initially encourage the children to keep their pieces short. After a brief discussion they may like to work for a little longer and develop some of the ideas that came out of the discussion.

◗ Workshop

◗ Performance and discussion

◑ Improvement time

Activity 4

Continuation of the original activity. Pairs now make groups of four, with similar sounding instruments.
* The children make up a different short piece, combining the sounds of all their instruments.
* Listen to the groups as before.

Advice

Some children will find it relatively easy to work together in groups in this way. Others will find the social aspect of the experience more difficult.

This pattern of working may need to be presented to the children on a number of occasions before it becomes an acceptable routine.

Quick whip

Aim

To create an instant class composition. The children will benefit if they have already worked on 'Take your partners please', or have some experience of grouping and sorting instrumental sounds.

Organization

Children and instruments in pre-arranged groups sit ready to watch the conductor and play their instruments.

These groupings and ideas are merely intended as suggestions and starting points. Each teacher must feel free to adapt and change these to suit his or her own circumstances and situations.

Group 1: Instruments made of metal
  chimebars
  triangles
  cymbals
  Indian bells
Group 2: Instruments made of wood
  woodblocks
  claves
  maracas
  guiros
  gato drums
Group 3: Pitched instruments
  xylophone
  glockenspiel
  metallophone
  recorder
  melodica
  violins
Group 4: Skin-covered instruments
  drums
  tambourine

Activity 1

Using the suggested conducting signs (see Figure 16) or others that you have devised:
1 Practise a '*tutti*' (everyone together) stop and start once or twice to ensure that you have everyone's attention and concentration.
2 Listen to each group in turn playing both loudly, softly; getting louder, getting softer.
3 Practise some of the other signs with the appropriate groups, i.e. trills/shakes, glissando.
4 Suggest that the children look carefully at their instruments and think of a way of making an interesting sound on it. Devise a conducting sign for 'interesting sounds'. Encourage the children to think carefully about playing loud or quiet. Practise a '*tutti*' loud and quiet.

Advice

There may be one or two children who seem unable to play their instrument and watch the conductor. This does require a certain degree of co-ordination and skill. Give these children a little extra practice.

Activity 2

Create a class piece. The following are intended as suggestions only:
* Conduct group 3 (pitched instruments) in and indicate that they should play a quiet continuous glissando.
* Conduct a brief, quiet 'conversation' between group 1 (metal instruments) and group 2 (wooden instruments) and end with a roll or shake from group 4 (the skin instruments).

Figure 16

Some useful conducting signs for class or group work

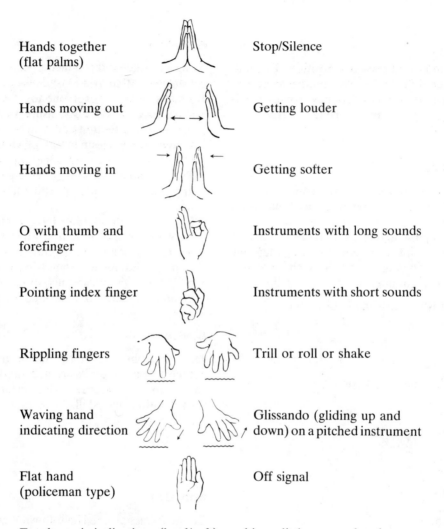

Hands together (flat palms)	Stop/Silence
Hands moving out	Getting louder
Hands moving in	Getting softer
O with thumb and forefinger	Instruments with long sounds
Pointing index finger	Instruments with short sounds
Rippling fingers	Trill or roll or shake
Waving hand indicating direction	Glissando (gliding up and down) on a pitched instrument
Flat hand (policeman type)	Off signal

For dynamic indications (loud/soft) use big or little arm or hand movements.

Always begin with a few at a time. Gradually build up a chart with illustrations.

These are just one or two suggestions. Encourage the children to devise some more.

Talking points

* 'What kind of sounds do the instruments in group 1 (or 2) make?' (If the children are confused about this, see 'How long does my sound last?' at the beginning of the book.)
* 'Did you like the piece? Would you like to change any part of it?'

Advice

It is always useful and helpful to tape record the pieces. In this way the children are able to take a more objective view of the whole piece.

Activity 3

* Ask a pupil to illustrate the sounds on the board as the sounds are made.

  You may start with:
  a roll on drums/tambourines (group 4)

  ∿∿∿∿∿∿∿∿∿

  3 long sounds (include the glockenspiels)

  o   o o

  4 short sounds (include the xylophones)

  ●●●   ●

  End with a roll getting louder, then softer.

  ∿∿∿∿∿~~~~

* Encourage children to come and conduct and create a piece.
* Discuss different ways of grouping the instruments.
* Display the graphic illustrations in the classroom.

Advice

Prepare some ideas for the shape of your piece beforehand. Keep the pieces short at first.

Give them a definite shape, i.e. begin and end in the same way or begin quietly and build the piece up group by group.

Discuss the pieces with the class.

What's in my piece?

Aim

Alongside the various class activities children should be working individually, in pairs, or in small groups on their own compositions. Opportunities should be given as often as possible.

Various approaches are possible with this kind of work. Here are some ideas as well as some suggestions and advice for helping children to develop and to shape their own compositions.

Possible activities

* Completely free approach: children select whichever instruments they wish and work on a piece. A number of ideas will emerge. Select one or two for discussion and use these to build on and to develop. (See 'Advice' below.)
* Children play selected groups of instruments (all wooden, all metal, all skin) to produce a piece focusing on only *long/short* sounds.
* Children focus their piece on a musical idea such as getting louder/quieter (dynamics); getting slower/faster (tempo/speed/pulse).
* Make up two rhythmic patterns. Play the patterns on different instruments. Combine them. Play them loud/quiet, slow/fast.
* Make up a piece with a musical surprise in it. Can the rest of the class guess what the surprise is?
* Select just a few instruments. Make up a piece showing a number of different ways of playing these instruments.

Talking points

Beginnings and endings:
* 'Did you all start together or one at a time?'
* 'Did you do something quite different?'
* 'Did you need a conductor? If so was it useful and did it work?'

How and where?:
* 'Was the music loud/soft, slow/fast, jerky/smooth, or none of these?'
* 'Did it get louder/softer, slower/faster?'
* 'Did anyone notice anything unusual/interesting/amusing about the piece?'
* 'Are you pleased with your piece?'

Advice

Whilst children are working in this way the teacher's role may need to be modified.

Listen/Look/Learn
* Observe and listen to group interaction and discussion as well as to the musical sounds being produced.
* Assist with care.
* Do they need any help?
* Is there a problem? What is it?
* Is it social or personal interaction?
* An instrumental difficulty? Technical?
* A musical difficulty (trying to cope with an idea that is beyond them)?
* Lack of motivation?
* Do they understand their task?
* Allow time for each group to perform their piece if they feel it is ready, and for brief discussion.
* Make careful suggestions about their compositions.
* It is always possible to build on ideas that they have initiated. Beware of imposing your adult ideas.

Group working

Performance and discussion

Da capo

Aim

Further ideas for helping children to structure and shape their own compositions, working from picture shapes all based on the musical sandwich idea – beginning and ending a piece in the same way with something tasty in the middle.

Organization and resources

Prepared workcards (see Figure 17 for suggestions).

A selection of instruments available, though some children may like to develop some of the ideas using their voices.

Children work in small groups.

Activity

* Each group chooses a card.
* The children should experiment, share and discuss their ideas.
* Encourage them to try different ways of representing the picture shapes. Is it a smooth shape? Is it a short detached shape? Does the sound they are making with their voices or on their instrument convey the idea behind the picture?
* The patterns show no indication of dynamics (loud/soft). Encourage the children to discuss this and decide where their piece will be loud or soft and how they could achieve this.
* Display all the cards. As each group performs their piece the others must guess which card they are interpreting.

How does it move?

Aim

To give teachers some suggestions and ideas for wording and preparing workcards to be used in the 'music area'.

Organization and resources

Prepared workcards as an initial stimulus. A wide variety of instruments available, particularly pitched or tuned instruments.

Activity

Figures 18–23 give examples of a series of self-explanatory 'workcards' using picture shapes.

The emphasis on most of the cards is on the idea of pitch, though children are encouraged to combine ideas of speed, pattern and dynamics (loud/soft).

Each card is worded and addressed to the children. They should now have the opportunity to spend as long as they need to develop their ideas.

Figure 17
Suggestions for workcards

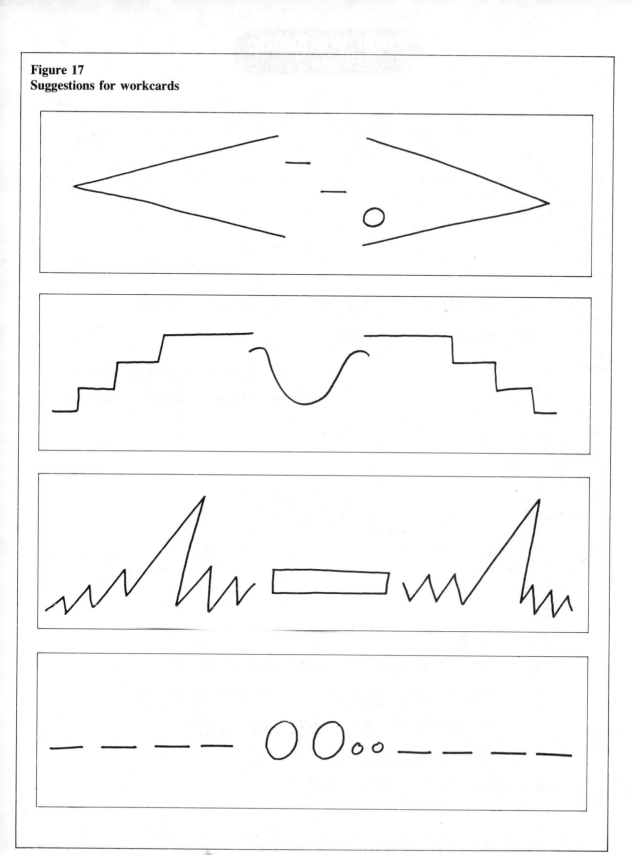

Figure 18

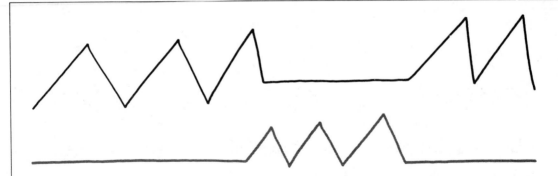

Use this pattern to make up a longer piece.
Think about the speed.

———— fast.... getting faster
———— slow.... getting slower

Figure 19

Play this pattern. REPEAT IT.
Now — play the pattern over and over again.
    Begin <u>very</u> <u>quietly</u>..... gradually make
    the patterns get <u>louder</u> <u>and</u> <u>louder</u>
    and <u>LOUDER</u>........... <
How does your piece end?

Figure 20

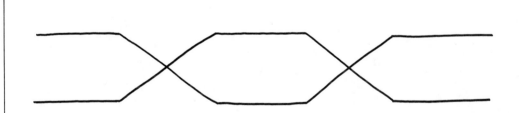

Play this pattern. Think carefully.... how do the shapes move up and down?

Now —   play the pattern
         <u>develop the pattern</u>
         return to the pattern.

Do you like your piece?

Figure 21

How do these patterns move up and down?
Use the pattern to make a longer piece.
Think carefully about the <u>beginning</u> } of your piece.
                          <u>ending</u>

Figure 22

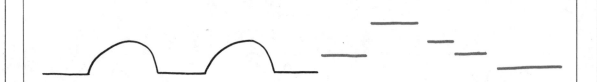

Use this pattern to make up a longer piece.
Play the pattern SLOWLY first.
Each time you play it get QUICKER.........
How does your piece end?

Figure 23

How does this pattern move up and down?
Play this pattern. CHANGE IT. Repeat the pattern.
Do you like your piece?

DIFFERENT
STARTING POINTS

This section is for children with some previous musical experience.

One enjoyable, rewarding and popular way of working in primary schools is to involve the children in a multi-media-based project or topic which may cover a number of days or weeks according to the way in which your school is organized. The starting points for these activities could be a variety of different sources: poems and stories, photographs and pictures, movement and drama, conversation and discussion, or a scientific or historical subject. The most obvious way of devising a 'musical' link in the topic work is to find a suitable song or piece of recorded music for either listening or movement activities.

On the following pages teachers will find suggestions for preparing and developing *workshop*-based activities resulting from three different starting points. When preparing and planning workshop activities it is important that teachers think of the *musical ideas* involved in the work that they hope the children will develop so that the results are *musical* rather than a series of sound effects.

These activities will always be more productive if the children have had both experience in groupwork, or pupil-initiated work in other curriculum areas and some previous musical experience. All the preceding activities and games in this book will help towards this experience; some of the activities will provide quite specific background experiences related to some of the topic ideas, and teachers are encouraged to work through some of them before expecting their children to work effectively in this way.

Ways of working
There will be a variety of opinions about the most effective way of beginning and then developing this work. A great deal will depend on the organization in individual classrooms.

A class discussion, where children pool and share ideas is often a useful springboard for subsequent group work. Teachers must ensure that they draw on and develop the ideas and suggestions made by the children and guard against the imposition of adult concepts and opinions in the pretence that this will provide the basis for a good model. There is a fine dividing line between providing enough stimulus, motivation and framework and restricting the children's imaginative and creative ideas.

When children are engaged in this kind of activity there is bound to be a period of indecision and slight confusion. There must be a time for selection and rejection as children experiment with ideas. It is impractical to prescribe a set pattern or time guide as teachers will and should devise ways of working appropriate to them and their children. However, although children must be given time and space to develop ideas they must not be allowed to flounder and become disinterested through lack of motivation or direction and help.

Teachers can leave experienced children to work for as long as they can unless they are invited to intervene. For less experienced children the following time guide might be useful:

Class introduction – 20 minutes

Group work – 15–20 minutes (for preparatory discussion and shaping of ideas.)

Class activity – 10–15 minutes (Recap on ideas. How far have we progressed? Listen to some ideas from one or two groups.)

Group work – 15–20 minutes. Further structuring and shaping of pieces.

Class activity – Performing, listening, discussing, recording.

The teacher's role

This will change as the activities progress. Teachers will advise, cajole and encourage. They must decide which groups need to be pushed gently further and how this might best be achieved and which groups need no teacher intervention. If problems arise they must decide on the nature of the problem. Is it social interaction, the personal make-up of the group; is it lack of enthusiasm or is it lack of knowledge or a skill that prevents the children from moving forward?

Contrasts 1

Starting point

Two poems with contrasting moods.

Aim

To create two contrasting 'sound pictures' using the poems as an initial stimulus.

Useful previous activities

'In the gap', 'How quiet?', 'Let's all hum', 'Take your partners please', 'What's in my piece?'.

Any other work and musical input concerned with dynamics and pitch.

Organization and resources

A variety of instruments and other sound sources should be made available.

After a class discussion children work in groups.

Poems either displayed or available on individual cards. Three poems are given as suggestions only.

Activity

* Read the poems.
* Do the children understand the vocabulary/the ideas/the images?
* How does the poet capture the atmosphere in each poem?
* How could that relate to musical ideas? The contrast of the very rhythmic qualities of 'The railway station' with the 'timeless' atmosphere created in 'Fog'.
* How will the children work? They could:
  1 speak the poems and work in musical ideas;
  2 use the contrasting atmospheres created by the words as a stimulus for musical contrasts – use ideas from some of the games and activities worked on previously;

  3 select some effective sound words from each poem and use these as a vocal accompaniment to the spoken poems;
  4 develop these sound words on their own; silent . . . drifts . . . soft . . . (see 'Poppp-BangggShShshsh');
  5 abandon the poems altogether and just develop the ideas of dynamic contrasts.

* Will the piece be instrumental, vocal or a combination of both?

Advice

Children must be given time to experiment, to discuss, to select and to discard ideas. The teacher should be on hand to encourage and to listen, to help organize practicalities and to help pupils to develop their ideas.

 Class discussion

 Workshop

Fog

Soft fog falls
Silent on the town.
It encloses everyone
On his small island.
All alone
Then drifts off.
Anon.

Fog

The fog comes
on little cat feet.
It sits looking
over harbour and city
on silent haunches
and then moves on.
Carl Sandburg

The railway station

Trains coming in,
Trains going out,
Buzzing, screeching,
Grinding, scraping.

Whizzing past you,
Not caring for you,
Whistles whistling,
Brakes screeching.

People running,
People walking,
Shuffling, scuffling,
Along the wide crowded platform.

People get on the trains,
Opening and shutting doors,
The platform is almost deserted,
People no longer exist there.
Gwyneth, aged 11

Pupil Workshop
Stage 2

Contrasts 2

Starting point

Instruments with contrasting timbres (sound qualities).

Aim

To compose a short piece of music to be called 'Contrasts'.

Useful previous activities

'Take your partners please', 'How quiet?', 'What's in my piece?'.

Organization and resources

A variety of instruments and sound sources. Children work in groups.

Activity

Children collect a group of instruments or sound-producing objects (not more than ten).
* As a group experiment with the instruments.

* Make sounds in different ways.
* Listen to the sounds each one is making.
* Talk about the sounds.
* Arrange the instruments in groups.
* Make up a pattern of sounds with one group of instruments and a contrasting pattern with another.
* Are the sounds *loud* or *soft*, *long* or *short*, *gentle* or *lively*? Are the patterns *fast* or *slow*, *jagged* or *smooth*? Think of other ways in which the sounds could contrast.
* Now the children use some of their ideas to make up a short piece.
* Let them practise their piece and play it to the other children.

Talking points

* 'How many different contrasts did the group use?'
* 'How were some of the contrasts achieved?'

Reflections

Starting point

Visual images.

Aim

To compose a piece of music based on the visual ideas of repeated and reflected patterns.

Useful previous activities

'Echo patterns' (rhythmic), 'Echo patterns' (melodic), 'How long is my sound?', 'Getting quieter', 'SSSSSS'. Song 'Long John'.

Organization and resources

A variety of visual stimuli: In the classroom – mirrors, spoons, foil papers, pictures, photographs, patterns. Outside – children collect data walking round the school, in the playground, in the street; windows, puddles, hubcaps, cars, water.
   Instruments available for group work.

Activity

* Let the children talk about the data and the information they have collected and observed.
* Let them share their ideas.
* Can they find words to describe the different kinds of images and patterns they have observed in the reflections? Repetition, reversal, symmetry, imitation, distortion – make sure the children understand these words in relation to what they have observed.
* Using felt-tips or paints they can draw patterns to illustrate some of these ideas. The patterns could be incorporated into 'picture scores' later. (See examples in Figure 24.)

* How will the children work? They could:
  1 build up a piece based on 'echo' patterns (both melodic and rhythmic);
  2 build up a piece based on the ideas of sound patterns getting quieter;
  3 using a picture or a photograph of reflections represent this in a 'sound picture';
  4 choose two or three of the visual patterns. Create a picture score and realize this in sound.

Advice

Some children will have their own ideas immediately. Give them time and space to work on these. Other children may need more structured ideas initially, while some may need help and advice in developing their own initial ideas.

Further suggestions for group work are given below.

Activity 2

Create a sound picture of a calm pond. Something happens to disturb this calm image. Eventually the water returns to its calm state. (See 'Da Capo' in the section *Developing Instrumental Activities*.)

Figure 24

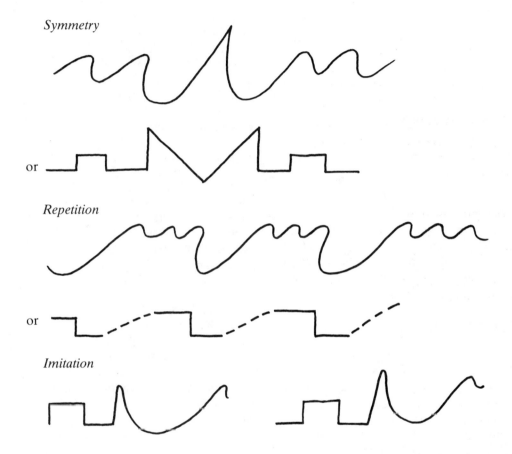

Symmetry

or

Repetition

or

Imitation

Activity 3

* Work on the idea of reflected distortions (hub-caps, spoons, water moving).
* Take a simple round, for example 'Frère Jacques' or 'London's Burning' – one that the children can pick out aurally. Experiment with the phrases. Change them. Play them backwards. Change some of the notes.

* In each of these rounds each phrase comes twice. Play one 'straight' and one 'distorted' in some way.

Activity 4

Take a vocal pattern and the musical idea of 'pitch'. Build up a piece that forms visual shapes, as suggested in Figure 25.

Figure 25(a)

'pitter patter'

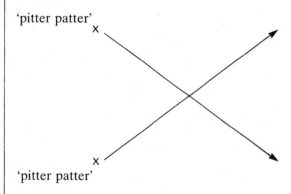

'pitter patter'

Working in two groups:
(a) one starts on a high note, one starts on a low note. Sing the word pattern, the 'high' group getting lower each time and the 'low' group getting higher each time. At some point the groups will cross over.
(b) both start on the same note and gradually move in opposite 'pitch' directions.

Figure 25(b)

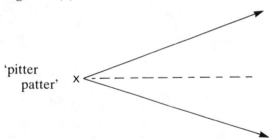

'pitter
   patter'

Work on two or three more ideas in the same way. Combine them into a 'piece'.

Activity 5

Take a poem, for example 'Water picture' by May Swenson. Represent some of these images in sound pictures.

Water picture

In the pond in the park
all things are doubled:
Long buildings hang and
wriggle gently. Chimneys
are bent legs bouncing
on clouds below. A flag
wags like a fishhook
down there in the sky.

The arched stone bridge
is an eye, with underlid
in the water. In its lens
dip crinkled heads with hats
that don't fall off. Dogs go by,
barking on their backs.
A baby, taken to feed the
ducks dangles upside-down,
a pink balloon for a buoy.

Treetops deploy a haze of
cherry bloom for roots,
where birds coast belly-up
in the glass bowl of a hill;
from its bottom a bunch
of peanut-munching children
is suspended by their
sneakers, waveringly.

A swan, with twin necks
forming the figure three,
steers between two dimpled
towers doubled. Fondly
hissing, she kisses herself,
and all the scene is troubled:
water-windows splinter,
tree-limbs tangle, the bridge
folds like a fan.
May Swenson

 or as long as the children are interested

Nobodyes and everybodyes gigge

Starting point

A historical project – 'The Elizabethan Age'.

Aim

To involve the whole class in a music-making activity connected with a 'history' project, combining individuals' expertise and beginners' enthusiasms.

Useful previous activities

'Patterns and pulses', 'Check your pulse'.

Organization and resources

Melody line of 'La Volta' written out and displayed (see page 80). Recorder players could prepare or practise this before the lesson.

Divide the class into four groups:

Group 1: melody instruments – recorders, xylophones, glockenspiels.
Group 2: bass instruments – bass xylophone bars, open-stringed cellos.
Group 3: percussion – tambours, tambourines.
Group 4: dancers.

Activity 1

* Set up a pulse of 3.
* Children count and tap the first beat in each group.
* Try this using different speeds. Make sure the children really feel the pattern of three beats.
* Decide on a suitable speed.
* Bring in the 'drone' accompaniment, as written playing every two bars (it is not necessary to write this out for the children).
* Add the melody line.
* Rehearse this. Play it loud then softly. Feel the dance-like quality of the rhythm.

* Invite one percussion player to fit a suitable rhythm pattern whilst the 'drone' and melody are played over and over again. The other percussion players join in.
* Once these three groups are playing, invite the dancers to experiment with simple steps that fit the music and that might be in keeping with the historical period, for example:

In pairs, holding hands – take
two steps forwards, (2 bars)
two steps back, (2 bars)
three steps forward and turn. (4 bars)

This could be danced in a long 'follow-my-leader' line or in lines facing each other. On the three steps forward and turn, couples will pass each other and turn ready to face each other and repeat the dance.

* Decide on an introduction and how many times the eight-bar phrase will be repeated.

La Volta

Melody

Drone

Percussion

or

Additional melody

Nobodyes and everybodyes gigge

Further suggestions for group work follow.

Activity 1

'Nobodyes Gigge' – Tell the children the story of the Elizabethan actor called Will Kemp, who is reputed to have danced from London to Norwich to the accompaniment of a pipe and tambour. (Kemp's own story, entitled 'Kemps Nine Daies Wonder' is available in the *Camden Society First Series*, Volume 2 (1840) available at Westminster City Library.)

*   In groups, pupils prepare their version of Will Kemp's jig.
*   Each group will need recorder players, percussion players, dancers.
*   Write out the melody on cards for the recorder players.
*   Let the children work out suitable dance steps to fit the music.

Activity 2

'Greensleeves' – Let the children work out their own suitable accompaniment – either melodic or just rhythmic – for this familiar melody played by recorder players. Guitar players would be quite useful.

Percussion

Recorder

Greensleeves

Class Activity
Stage 3

Nobodyes and everybodyes gigge

Activity 1

'Shapes and patterns' – A possible 'follow up' to work done on the architecture of the historical period, particularly the beautiful shapes and patterns used in the vaulting in churches.

* Sing the hymn 'Praise him from whom all blessings flow' (Tallis Canon).
* Divide into two groups and sing this as a round or 'canon'.
* Once the children are very familiar with the melody divide the class into four groups.

Tallis Canon

(Entry either at ① or ② for canon.)

Advice

Encourage the groups to listen to each other as they sing and to the strands of melody weaving in and out of each other.

Include some recorder players in each group to help keep the pitch.

Activity 2

Either working individually or in small groups, children can trace the shape of the melody.
* Using squared paper plot the almost step-wise movements of each phrase, as shown in Figure 26(a).

* Using different colours for each 'entry' they can show how the melodies weave in and out when the hymn is sung in 'canon'. Figure 26(b) illustrates this.

Advice

You may need to sing the melody through on two or three different occasions before the class can successfully attempt it in four parts.

Figure 26

(a)

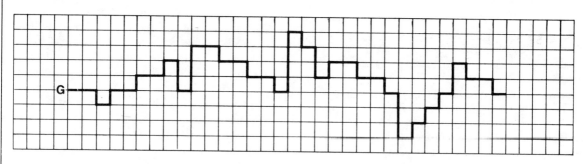

(b)

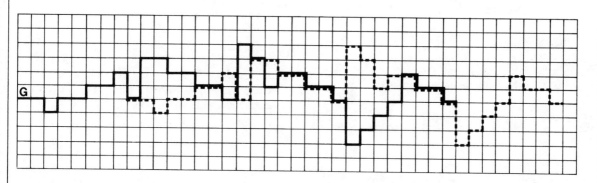

Nobodyes and everybodyes gigge

Activity 1

'Shapes and patterns' – Each group will need players for the melody line, players for the bass line, and players for the percussion.

* Write out the musical extracts, shown in Figure 27, onto cards.
* Set up a pulse of 3.
* Children count and tap the first beat in each group.
* Decide on a suitable speed.
* Work out the bass line. Use bass xylophone bars, a bass recorder (if you are lucky enough to have one) or find the notes on the piano. The same bass line has also been written in the treble clef. It may be easier for the children to read.

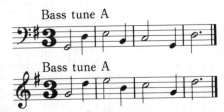

* Now the melody-line players work out their tune. Fit this above the bass line.

Melody

* Percussion players add some rhythmic accompaniment. Practise this a few times.

Figure 27   Rhythmic accompaniment

* Now play the bass tune shown in Figure 28(a). It begins like the first bass tune but does something different at the end. The same tune has also been written in the treble clef in (b).

Figure 28

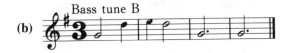

* Divide the group in two: Group 1 sings bass tune A; Group 2 replies with bass tune B. Repeat this once or twice. These two bass tunes are like a question and answer.
* Work out a melody line to fit above the four-bar answer tune B.
* Now play the first four-bar melody and bass tune and follow it with the answering four-bar tunes.
* Add some percussion accompaniment to this eight-bar piece.
* Practise it. Think about the dynamics. Play it through loud, then softly. Try it at different speeds.
* When they are ready let the children perform their versions to each other.

Advice

When children are fitting two or more parts together it is important that they feel the pulse and count carefully so that they come in together.

PENTATONIC PATTERNS

This section contains melodic activities for children with some previous musical experience.

Pentatonic is by definition five notes. A pentatonic scale can be any five notes. The two scales used here are the most useful and widely used in the primary classroom (see Figure 29), especially since Kodály's and Carl Orff's influence. The pentatonic scale provides us with one useful way of developing rhythmic and melodic ideas. However, teachers must remember that although it is a useful and convenient tool, children's melodic experiences should not be restricted to these intervals alone. Teachers must ensure that their pupils have adequate opportunities for singing, listening to and creating in a wide variety of other melodic styles.

A great number of songs, particularly folk tunes and spirituals, are based on pentatonic scales. Two songs – a folk tune and a spiritual – taught as a class activity provide the framework for a number of pupil workshop activities in this section.

Teachers are advised to prepare this work carefully and to familiarize themselves with melodic phrases before embarking on the vocal work with their class. Suggestions for doing this are presented with the activities. Two-bar phrases are demonstrated, as in the rhythmic work. This is musically much more appropriate but initially they ought to be very simple.

Figure 29

Scale based on C

C D E G A C'

Scale based on G

G A B D' E' E D

Scale based on D – for trumpets and clarinets when other instruments are using the C-based scale

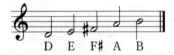

D E F♯ A B

My turn/your turn (melodic)

Aim

Further work to help develop melodic vocabulary using pentatonic scales.

Useful previous activities

'My turn, your turn', 'Singing names 1 and 2' (both in section *Vocal Activities*).

Organization and resources

Prepare melodic phrases using notes from the C-based scale, as suggested in Figure 30. Children sit comfortably in a group. The leader stands in front.

Activity 1

* Adam sings the pentatonic scale up and down to 'lah' (Figure 31).
* Jean could play it with him on the alto xylophone.
* The others join in.
* Decide on the tempo (speed).

* Demonstrate the length of the phrases – two bars.
* Using 'My turn/your turn' idea Mary sings a melodic phrase.
* The class sings the phrase back immediately.
* The activity proceeds in this way for two or three minutes without stopping.
* Encourage the children to use a variety of dynamics – where is it to be loud or soft?
* Make sure the class echo the phrases precisely.
* Repeat the activity, inviting different pupils to take the lead.

Advice

Initially the teacher will have to lead this activity but children should be invited to take the lead as soon as possible even if they only manage one or two phrases.

You may need to revise the pitch accuracy each time a different leader begins, by singing up and down the notes of the scale altogether.

The 'echo' activity should proceed unaccompanied.

Use an alto xylophone as a melodic reference point only.

Figure 30

Figure 31

Activity 2

Repeat the activity, choosing phrases based on different pulses and speeds. Suggestions are given in Figure 32. Demonstrate the pulse and the speed each time.

Figure 32

* Peter decides on a pulse and a speed.
* He demonstrates this to the class.
* He sings a phrase.
* The others echo as before.
* He continues the singing and echoing with one or two more phrases.
* Then he repeats exactly the *same* phrases with exactly the *same* dynamics.

Advice

Although suggestions are given here, teachers are encouraged to prepare their own vocabulary of phrases.

Always establish the speed and the pulse before singing the phrases. The speed and pulse will remain constant during each leader's 'turn'.

Activity 3

* Proceed as before.

* If he does this the class do *not* echo it back.
* Repeat the 'game' with different leaders.

Advice

This is a more difficult activity to lead. Pupils will need some experience to do this effectively.

Vary the phonics to sing to 'lah' or 'doo-doo' etc.

Activity 4

Repeat all these activities using phrases built up round the G-based pentatonic scale (shown at the beginning of this section). Suggestions are given in Figure 33.

Figure 33

Advice

It is important for teachers to use the correct vocabulary, i.e. 'unison', 'dynamics', but constantly check that the children understand and remember what the words mean. Refer to relevant musical examples as they arise in the children's pieces.

Pentatonic magic

Aim

An introduction to some simple melodic accompaniments, using some of the C-based patterns previously introduced.

Organization

Song 'Hop tu nay' – a playground chant from the Isle of Man. (See page 91 for words and melody line.)

Children sit comfortably in a group.

Activity

* Sing the song through, unaccompanied, to the class.
* Teach the song to the class.
* Extract one or two rhythmic phrases from the song.
* Clap these to the class one at a time:

Hop   tu   nay

Mo - ther's gone a-way

Gin-ny the witch

* Can anyone match these to the correct word phrases?
* Can anyone sing them?
* Choose one pattern. Let everyone clap this over and over:

* Select a small group to continue clapping the rhythm pattern whilst the rest of the class sings the song through. The accompaniment group could clap the pattern twice through as an introduction before the singers come in.
* Select another pattern:

Hop   tu   nay

* Again, all the children clap this through a few times. Select another small group to sing this as a second accompaniment pattern.

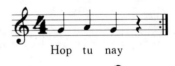

Hop tu nay

* Make sure the children know which group will come in first.
* How will you decide to end?

Talking points

* 'Can anyone comment on the shape of the song?' (See 'Da Capo' in the section *Developing Instrumental Activities*.)
* 'What pulse pattern is used in the song?'
* 'Can anyone clap this pulse pattern?'

Advice

This chant also works well as a round, but it is advisable not to make it too complicated at first.

Hop tu nay

Hop tu nay, My mother's gone a-way, And she won't be back un - til the

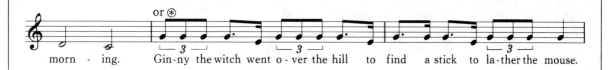

morn - ing. Gin-ny the witch went o - ver the hill to find a stick to la-ther the mouse.

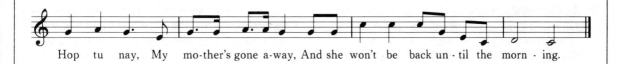

Hop tu nay, My mo-ther's gone a-way, And she won't be back un - til the morn - ing.

Pentatonic magic

Aim

To give the children the opportunity to develop these ideas from the song 'Hop tu nay' in their own way.

Organization and resources

Children work in groups – six would be quite useful – with pitched instruments available for each group; one or two non-pitched instruments; C-based pentatonic patterns on cards.

Activity

* As a group sing through the song. Although we are using C-based patterns the actual melody of the song begins on the note G. You may need to play this on a xylophone so the children all start on the same note.
* Which pulse group are you working with?
* Mandy works out a melodic pattern based on the steady pulse, for example:

* Peter makes up a rhythmic pattern on the woodblock:

* Saleem could use a melodic phrase from the song:

* James, Wendy and Ian sing the song through whilst the others try to fit in the accompaniments.
* Decide in which order the patterns will build up. Who will come in first? Will you use the patterns as an introduction? How many times will you sing it through?

Advice

When children are fitting two or more parts or patterns together it is important that they feel the pulse and count carefully so that everyone recognizes where the first beat of each pattern is.

Make use of children learning other instruments such as violins, cellos and flutes.

Beginner trumpets and clarinets could play phrases using a C-based scale. Write their phrases on a D-based scale (see Figure 34 for suggestions).

Figure 34

More magic

Aim

Further work on melodic and rhythmic accompaniment patterns, using G-based phrases.

Organization and resources

Children sit comfortably in a group; instruments available if required.

Song 'Standin' in the need of prayer' – a spiritual. The arrangement on page 94 is presented as a suggestion only. Teachers are encouraged to develop their own and the children's ideas.

Activity 1

* Sing a chorus and a verse of the song to the class, unaccompanied.
* Sing the chorus through again, indicating the beats of '1' and '3' with a finger click.
* Invite the class to join in with this beat.
* Sing the verse through and ask the class to be ready to come in with the chorus.
* Repeat this.
* The class should now be ready to learn the verse in the same way.

Advice

The song begins on an anacrusis (an 'up' beat). It is important that the children are aware of this because the accompaniment patterns will start on the first beat of the pulse group.

Care needs to be taken in teaching the semitone phrases in the verse. Play some melodic games using semitones. The G-based scale is easier for descant recorder players, as the low C can cause difficulties.

Activity 2

* Sing the following pattern to a suitable phonic sound over and over again:

* Invite the class to join in.
* Choose a small group who will keep this going as an accompaniment. They start; then bring the rest of the class in on the last beat of the pattern. They sing through the chorus and a verse.
* Introduce one or two more G-based phrases in the same way. Add these to the accompaniment:

* Make sure that everyone knows in which order they will come in.
* Build up a performance of the complete song.
* Vary the pattern combinations for different verses.
* Which patterns will be most effective as an introduction?
* How will you link the verses?
* The rhythm of some patterns could be played on unpitched instruments.

Advice

Keep the initial combinations of patterns fairly simple and use just one or two straightforward patterns to begin with. As the children become more familiar with the song more difficult combinations can be introduced.

Standin' in the need of prayer

arr. L. Davies

Vocal

It's me, it's me oh Lord,

(do do da da)

Xylophone or Glock.

Woodblock

Stan - din' in the need of prayer. It's me, it's

me oh Lord, Stan - din' in the need of prayer. Not my

Fine

[94]

Pentatonic patterns

Aim

To allow children the opportunity to use and to develop some of the ideas covered in the class activity and to make up a short piece combining some of these ideas with some of their own.

Organization and resources

Children work in small groups.

Make available, in sets, the notes of each pentatonic scale on a variety of pitched instruments.

Prepare cards with some melodic and rhythmic suggestions based on ideas already worked on in the class activity earlier in this section (such as the examples given in Figure 35).

Activity

* As a group the children decide on a speed.
* Each group has a set of cards. Together count out the pulse indicated on the card at the agreed speed.
* Work out the prepared patterns.
* Play these together.
* Try playing them at different speeds. Is is better faster or slower?
* Sarah and Melanie play one phrase. Peter and Stavros follow on with the second phrase.
* Using the notes of the scale the children can experiment individually and make up some more two-bar phrases. Use these with the two prepared phrases to make up a short piece.
* Peter could play his phrase over and over again (an ostinato), while the others play their phrases in turn over this.
* Let the groups perform their pieces.

Talking points

* How did the group organize their patterns?
* Did they play each pattern in turn, in unison or did they combine some patterns?
* Did they make any use of dynamics?
* How did the piece finish?

Figure 35

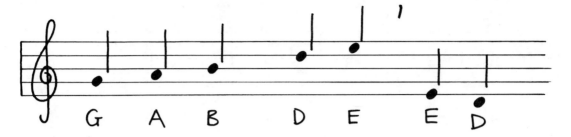

G A B D E E D

Make up some patterns using some of these notes.

A walking pattern
count a steady 4 beats

A running pattern with a long note to end.
count a steady 4 beats

Figure 35 (contd)

A 'jazzy' pattern with a hiccup at the end.

A pattern using long notes.

Count a steady 4 beats

1   2   3   4

Choose some more notes and make up your own patterns.

A friend could play the 1 2 3 4 pulse group on a drum. Can you fit your pattern over this pulse ?

SHALL WE SING?

In this section a number of songs have been used to provide a framework and stimulus for most of the activities.

Singing is probably the most popular and widely experienced source of children's music making and although no set pattern for teaching singing is prescribed here, the class activities encourage teachers to take a fresh look at the way in which they prepare and present song material. Teachers are also encouraged to approach the teaching of singing as an enjoyable *skill* which must be worked on in the same way that enjoyable instrumental skills are approached and worked on. Children should listen to themselves and each other; they should sing in 'musical sentences'; to breathe at the end of phrases; to be aware of and to articulate the words of songs with care.

Five specific songs are presented to give teachers some suggestions and ideas for building up and using rhythmic and melodic accompaniment patterns for songs. The workshop activities suggest ways of developing some of the musical ideas used in the songs. The whole class is able to participate and be involved in this enjoyable musical experience even though all these activities require a degree of musical understanding on the part of the teacher and the pupils.

General points on accompaniment

Accompanying a song can be a fairly sophisticated musical activity. It involves fitting together any number of rhythmic and melodic patterns. There is always the danger that a song which the children sing well may deteriorate with the addition of even a few percussion instruments if the children are unused to hearing different sounds and patterns happening at the same time. The more general musical experience your children have the more successful your accompaniments will be.

The simplest way of adding a rhythmic accompaniment is to play the pulse beat of the song:

Children should experience this through clapping or moving whilst they are singing. There are many early rhythmic activities and games in the book which will help the children to develop this skill.

Rhythmic ostinati

These consist of short rhythm patterns for percussion – or body sounds – usually one or two bars in length, which can be repeated throughout the song or sections of it. Many of these phrases are taken from the rhythm of words or phrases in the text, for example:

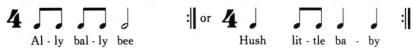

It will help the children if they say the words as they play, at first. Let them experiment with finding their own word patterns from the songs.

Melodic ostinati or second melody lines

In the early stages try to keep these simple. It always helps if there is some kind of pattern or repeated phrase that the children can *memorize*. Playing a xylophone or glockenspiel and keeping an eye on music is quite difficult.

Devising an introduction

Always make sure the children know what the introduction is. Use some of the rhythmic or melodic phrases, but make sure the children know how many times they will be played before they begin to sing. Make sure the children know on which note they start singing.

General points

Piano accompaniments have not been provided here. Children will learn the melodies much more quickly and accurately if the songs are sung unaccompanied. They can then really hear the sound of their own voices. If you prefer the support of an instrument to singing on your own, the guitar is a good alternative to the piano and chords have been provided. In general never let the accompaniments dominate the singing. Take care with the dynamics (variations in loud and quiet). Children have a tendency to sing and to play everything at the same volume and they need help and reminders if they are really to appreciate these contrasts. Indications of dynamics and speed have been made but always discuss these with the children and let them decide on the treatment which they feel is appropriate to the character of the song. Attention to this sort of detail is an important part of the children's musical development and will help to make a much more musical performance.

Using the patterns

Five songs are presented here to show: (a) how to devise and prepare accompaniments, and (b) how to work on these with the children. Teachers are encouraged not only to modify these to suit their own pupils and resources, but also to examine the way these patterns have been devised so that they can develop the ideas for further song material.

The children must be familiar with the songs before the accompaniments are introduced, particularly the melodic ones. Make sure the patterns have been made available in the music area for individual practice. Allow as many children as possible to have a turn at playing an instrument during the course of a lesson. Use different instruments for different verses. Discuss the suitability of instruments with the children. Make sure a tempo and a pulse have been clearly understood. Always sing through at least one verse of the song altogether before adding any accompaniments. Introduce the simplest pattern first (these have been numbered in order of difficulty). Avoid introducing too many patterns at one time. If problems arise, return to the simplest patterns – these are often the most effective – and allow further opportunity for the children to experiment and practise in the music area. No 'stages' have been suggested for these activities as they can be made as simple or as complex as you wish or as the pupils' skills and resources allow. All the songs are suitable for junior children of any age to sing and enjoy.

Ally Bally Bee

One simple melodic line consisting of minim notes. Make sure the children hear this against the crotchet beats in the melody line. A few rhythmic patterns based on 'word patterns' are also given, but the children should be encouraged to devise their own. Children should suggest suitable instruments.

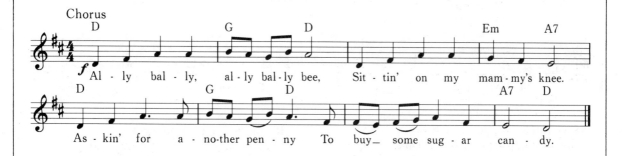

Little Annie lookin' very thin,
A couple o' bones covered o'er with skin,
Now she's got a little double chin
From eatin' sugar candy.

Chorus

Mammy gi'me a penny do,
Here's old Coulter comin' round,
Wi' a basket on his crown,
And sellin' sugar candy.

Chorus

Don't you cry, my bonny wee babe,
You know your daddy's gone to sea,
Earnin' pennies for you and me
To buy some sugar candy.

Rhythmic accompaniments

This accompaniment works well on glocks, metallophones or recorders.

Hush Little Baby

Two melodic patterns which will work well on glockenspiels or recorders are suggested. Introduce them one at a time. An interlude between the verses is also suggested. Ask the children to suggest suitable instruments and rhythmic patterns. It would be sensible to have different instruments for each verse. Decide whether the interlude is more effective between each verse or every two verses. A simple pulse beat can be very effective in this song.

1. Hush lit-tle ba-by, don't say a word, Ma-ma's gon-na buy you a mock-ing bird.

2 If that mocking bird don't sing,
Mama's gonna buy you a diamond ring.

3 If that diamond turns to brass,
Mama's gonna buy you a looking glass.

4 If that looking glass gets broke,
Mama's gonna buy you a billy goat.

5 If that billy goat won't pull,
Mama's gonna buy you a cart and bull.

6 If that cart and bull turn over,
Mama's gonna buy you a dog named Rover.

7 If that dog named Rover won't bark,
Mama's gonna buy you a horse and cart.

An interlude

Use this interlude between every pair of verses – play on glocks, xylophones or recorders.

This is a very simple song which can be made interesting with careful accompaniment. Choose 'quiet' sounding instruments. Ask the children for suggestions. Use a different instrument for each verse. Keep a simple pulse beat, or just a beat at the beginning of each bar.

Here are two melodic patterns which work well on glockenspiels or recorders. Print them out onto card for the music corner so that every one can practise them.

The Soal Cake Carol

This is a very simple three-note melody which can be made as simple or as complex as your skills and resources allow. Introduce each melodic pattern gradually. Pattern 2 can be a vocal pattern for a small group or it can later be extended to provide a quite different counter-melody. The main verse also works well as a round.

'God rest you merry gentlemen' could be added (in the key of E minor) to make this a very sophisticated performance. In order to attempt all these suggestions you may need a larger number of children than just one class.

Patterns

Melody I Bright and fast

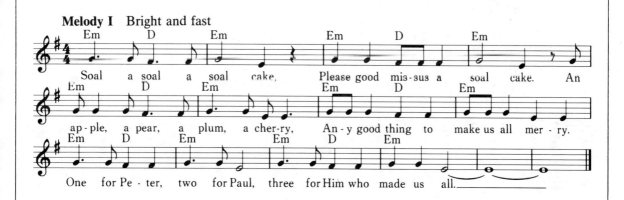

Melody II

Hey - ho, no - bo - dy home. Meat nor drink nor mo - ney have I none.

Yet shall we be mer - - - - - ry.___ Hey - ho, no - bo - dy home.

Verse

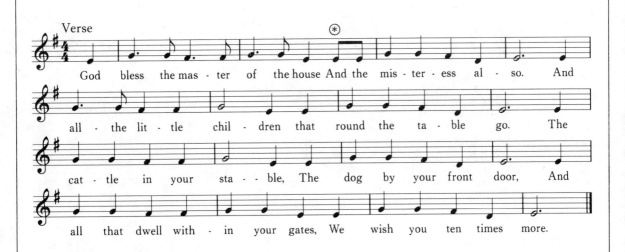

God bless the mas - ter of the house And the mis - ter - ess al - so. And

all the lit - tle chil - dren that round the ta - ble go. The

cat - tle in your sta - ble, The dog by your front door, And

all that dwell with - in your gates, We wish you ten times more.

Ladybird

The melody and rhythm of this song are quite tricky and require careful teaching before any accompaniments are introduced.

The pitched accompaniments are particularly designed for open strings, although chimebars or xylophones can be substituted. Use the first drone (A and D) as a two-bar introduction and an interlude between verses. Experiment with the accompaniments 1, 2 and 3 – try them separately and together with each of the verses.

The words are based on the children's rhyme – if the ladybird stayed on your hand then it would grant you a wish.

Gentle but not too slow

by Ted Edwards

1. La - dy - bird sat on me 'and I know if tha stays till I've fin - ished me rhyme What - ev - er it is I com - mand, What - ev - er I'm wish - in' for soon will be mine.— In our 'ouse there's no din I wish there were some - bo - dy in. La - dy - bird, la - dy - bird, fly a - way wom, Thi 'ouse is on fire and thi' chil - der are gone.—

2. Ladybird, listen to me.
   I've not seen my fayther for many a day.
   He comes wom whenever he's free,
   But not very much 'cos he works far away.
   Last year he worked in the town.
   Now they're pullin' the factory down.
   Ladybird, ladybird, fly away wom . . .

3. Ladybird, in't it a crime?
   Me mam is a-workin' away deyn in t'mill.
   'Er'll come wom about supper time.
   'Er's on afternoons, so I've hours to kill.
   I'll queue for some chips and some fish.
   Then I'll switch on the telly and wish.
   Ladybird, ladybird, fly away wom . . .

Accompaniments

For chime bars or violins - open strings

Chimebars

Violins

The drones for chimebars/violins can be played on bass or
alto xylophones - split the chords into this pattern:

This will also work in the following way for cellas on open strings

Side drum

Add a roll in the last 8 bars

(Chorus)

Land of the Silver Birch

This song is based on the G pentatonic scale (see section on *Pentatonic Patterns*). Allow the children to work out for themselves both the rhythmic and melodic patterns which might be suitable. They might like to find ways of writing these down. One or two are suggested, however, in the section on *Pentatonic Patterns*. This song also works effectively as a round.

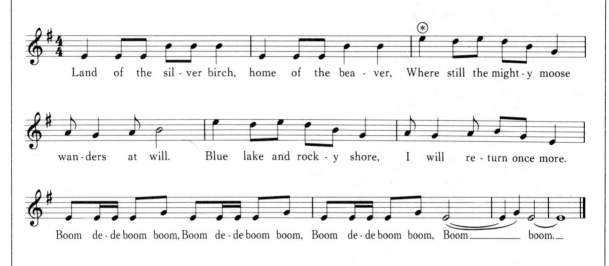

Land of the sil-ver birch, home of the bea-ver, Where still the might-y moose

wan-ders at will. Blue lake and rock-y shore, I will re-turn once more.

Boom de-de boom boom, Boom de-de boom boom, Boom de-de boom boom, Boom boom.

Pattern 1

Repeat throughout.
Use xylophones – alto or bass produce the best sounds.

Pattern 2

Repeat throughout.
Use glocks or recorders.

Building the patterns

Aim

1 By providing some initial examples of both rhythmic and melodic accompaniments to songs, teachers will find ways of developing these ideas to devise accompaniments for further song material.
2 To suggest ways in which pupils can use, practise and develop these ideas as workshop activities.

Organization and resources

Teachers will need to familiarize themselves with both the songs and the suggested accompaniment patterns before they present the material to the children.

The melodic and rhythmic patterns should be written onto prepared workcards. A variety of pitched and unpitched instruments should be available in the music area.

Activity

As each song is presented as a class activity, the appropriate prepared workcards should be available in the music area.

1 Children can work *individually* – for practising the patterns.
2 Children can work *in pairs* – for practising and combining the patterns.
3 Children can work *in groups* – for using and developing the ideas and working out their own versions of the songs. This is a sophisticated activity and the children will need to work on it over a period of time.

Major/minor: 'Long John'

Aim

Through singing, playing and listening the children will become familiar with the different qualities of the major and minor harmonies.

Organization and resources

Prepared song (see page 112); children sitting comfortably, ready to sing.

Activity 1

* Warm up with 'My turn/your turn (rhythm)' (in the section *Rhythm, Pulse and Tempo*).
* Jason establishes a pulse of 4 beats at a moderate speed.
* He claps a pattern which the children immediately echo back:

* Jason continues in this way for two or three minutes.
* Change to a different leader if you wish to continue the activity.
* Play one round of 'Names on the click' (in the section *Rhythm, Pulse and Tempo*).
* Choose a small group to establish and to maintain this pattern at a moderate speed:

4 – – – x – – – x – – – x

* Choosing some of the rhythmic patterns from the song play 'My turn/your turn' (the teacher will need to prepare these), each pattern will begin on the 'click' which the pulse group is maintaining (see Figure 36).
* Teach the song using the 'My turn/your turn' idea.
* Establish the pulse, using the small group to maintain it. Sing the first phrase (as shown in Figure 37).

Figure 36

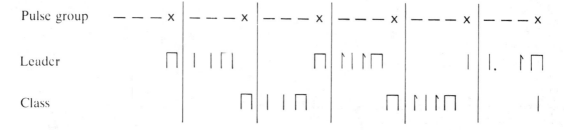

Figure 37

etc.

* The class echo this back.
* Sing and teach the rest of the song in the same way.
* Repeat this, singing through the song once or twice more.

Talking points

* 'What do you notice about the way each phrase begins?'
* 'Can anyone hear which phrases in the song have some notes which change?'

Advice

The song begins on the last beat of the bar – on an anacrusis. By teaching the song in this way the children should have little difficulty in understanding this rhythmic feature.

Notice the dynamic markings in the song. Where is it loud or soft? Establish these as you teach the song – they are very important to the character of the song. Make sure the children echo the phrases accurately.

Long John

Major / minor: 'Long John'

Aim

Having learned the song the children can work out some of the phrases by ear and play them on a pitched instrument. This will help to reinforce the aural notions of the different qualities of the major and minor harmonies.

Organization

Children work in pairs. Chromatic pitched instruments and two beaters each available.

Activity

* Working individually, Saleem and Sarah will pick out some of the phrases from the song.
* Let them sing or hum the phrases as they work them out.
* Can they work out the phrases where one of the notes changes?

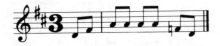

* Saleem decides to be the leader.
* Using some of the phrases they have worked out they play 'My turn/your turn'.
* How is Saleem playing the phrase? Is it loud or quiet? Sarah must try to echo this accurately.

Advice

On an alto xylophone the melody will be played an octave higher than as written in the song.

It may be necessary or helpful to tell the children which note the song begins on, though some will find their own pitch and key. If they work it out in a different key this could provide an interesting discussion point.

Can they sing the phrases in the key that they may have worked out?

Major/minor: Take a chord

Aim

Further work on major and minor harmonies.

Organization and resources

Children working in pairs or small groups, with a variety of pitched chromatic instruments and recorders. Prepared cards A and B (see Figure 38).

Figure 38

Activity 1

* Using cards A and B, Mary and Nazreen play the notes on card A; then they play the notes on card B.
* Can they hear which of the five notes is different?
* Let them experiment with the series of notes: play them quickly and slowly; play them in different orders; listen to the difference as they play.
* Mary makes up a short tune using the notes contained on card A. Nazreen does the same using notes from card B. (The scale on card A makes a major sound. The scale on card B makes a minor sound.)

Activity 2

* Marvin and Senga use cards A and B. They will need either a keyboard instrument, a chromatic xylophone or a glockenspiel.
* They play the notes marked 1, 3 and 5 on both cards. They must listen carefully to the sounds.
* They can play the notes together or one after the other.
* Can they make up some patterns using the three notes on card A and some patterns using the three notes on card B? (The three notes on card A make up a major chord. The three notes on card B make up a minor chord.)

Activity 3

* Using a combination of the three notes of the major chord, card A, Marvin makes up a pulse pattern based on a steady count of three beats, for example:

Figure 39

* This could be an accompaniment for a 'waltz' tune.
* Senga makes up a waltz tune to go with the accompaniment, using the same three notes.
* Marvin and Senga fit their tune and accompaniment together. It might be easier to begin with if the accompaniment pattern begins. A suggested tune is given in Figure 39.

Activity 4

Zung and Thuy prepare a tune and accompaniment in the same way using a combination of the three notes of the minor chord, card B.

Activity 5

Marvin and Senga and Zung and Thuy can use some of the ideas from their tunes and accompaniments to make up a 'question and answer' musical conversation.

Major/minor: Take a tune

Aim

Further work using major and minor scales and tunes.

Organization and resources

Children work in pairs or small groups. They will need a variety of pitched chromatic instruments and recorders; prepared cards C and D (see Figure 38).

Activity

* Tu Co and Jason play through card C on any pitched instrument.
* Now they play through card D.
* How does the tune change? Can they find out how many notes in tune C are different in tune D?
* Let them experiment with the tunes. How many different ways can they play them (loud/quiet; slow/fast).
* Play 'My turn/your turn'. Tu Co plays a phrase from tune C. Jason replies playing a phrase from tune D.
* Can they make up a 'major minor' tune using phrases and patterns from both tunes?
* Let them perform their piece.

Advice

In this tune a phrase would normally be either two or four bars, but do not worry if the children have a different idea.

Is a song just a song?

Aim

A different approach to learning a song, by extracting musical elements from the song and using these in activities and workshop situations.

Organization and resources

Children sit comfortably in a group.

The teacher must learn the complete song 'Smuggler's Lullaby' and prepare melodic and rhythmic phrases from it.

Activity 1

* Begin with 'My turn/your turn'. Choose a pulse of four. Echo a variety of rhythmic patterns using body sounds.
* Introduce some of the rhythmic patterns from the song:

* Play 'My turn/your turn' with melodic patterns in the same way, using voices. Choose a suitable sound to sing.
* Make up patterns over a range of these notes. You may need to prepare these beforehand.
* Introduce some of the patterns from the song (see Figure 40).

Figure 40

Smuggler's Lullaby

Manx Traditional

See— the— ex - cise men are com - ing, Sleep my— lit - tle he - ro,

They'll be— seek - ing wine and whis - key, Sleep my— lit - tle he - ro.

Och yene llia - noo mein cha - dil— oo ma lay la.

Och yene llia - noo mein, cha - dil— oo ma lay la.

Chorus
Och yene llianoo mein,
Chadil oo ma lay la. (repeat)

2 Daddy's late and we must warn him,
Sleep my little hero,
This run he'll have not illegal,
Sleep my little hero.

Chorus

3 Oh, the English men may board her,
Sleep my little hero,
Nothing wrong will they discover,
Sleep my little hero.

Chorus

4 Let them search in boat or dwelling,
Sleep my little hero,
Nothing's in the hold but herring,
Sleep my little hero.

Chorus

Talking points

* 'Can anyone pair up the rhythmic patterns with
the melodic patterns?'
* 'Which melodic patterns move up or down?'
* 'How do they move up or down?'

Activity 2

* Sing the complete song to the class.
* Teach the song to the class.

Advice

As the children are already familiar with some of the musical elements from the song they should learn this quite quickly.

Activity 3

* Take four particular melodic elements from the song (see Figure 41).

Figure 41

* Teach these, using voices, to the whole class before dividing into four groups.
* Subdivide Group 1 and establish the 'drone' effect. Whilst Group 1 maintains this, introduce the other patterns one by one.
* Using the four patterns make up a class piece.

Talking points

* 'Which pattern shall we start with?'
* 'How many times shall each group sing their pattern?'
* 'How will our piece finish?'
* 'Do we need to put a plan on the board to remind us of what we are doing?'
* 'What about the dynamics?'

Advice

Introduce the patterns slowly. Encourage the children to sing accurately. If they seem to have difficulty in holding the pitch use just the 'drone' by playing the two notes on a xylophone as they sing.

Activity 4

Work in the same way using instrumental groups instead of vocal groups.

* Write out each pattern clearly on large manuscript paper.
* Suggested instrumentation: recorders; xylophones; glockenspiels; metallophones; chimebars. Open strings on violins and cellos will provide a very satisfactory 'drone'.
* Make up a class piece as before.
* Tape record the piece.
* Can the children pick out their own parts?

Is a song just a song?

Aim

To reinforce the class activities and allow children scope for individual experimentation and interpretation.

Organization

Children work in groups, ideally four to six children in a group.

Each child has an instrument. Use both pitched and unpitched in each group. Four prepared cards for each group.

Activity

* Children experiment freely with the melodic patterns.
* Choose two patterns and play the rhythms on unpitched instruments.
* Using all or just one or two of the patterns let each group make up a short piece.
* Let each group perform their piece.

Talking points

Pick out one or two features from each piece to question the others about.
* 'How did the piece begin or end?'
* 'How did the group organize their patterns?'
* 'Did they use all the patterns?'
* 'Which instrument played Pattern 2?'
* 'Did they use dynamics in any special way?'

Advice

To begin with each group's ideas and organization may be very similar. The important thing is that each child has been able to make an individual contribution.

After the discussion time the groups may wish to improve their pieces and work further on them.

 Discussion and performance

 Improvement time

WHERE HAVE WE GOT TO?

Significant learning combines the logical and the intuitive, the intellect and the feeling, the concept and the experience, the idea and the meaning. When we learn in that way we are whole . . .

From *Freedom to Learn in the Eighties* by Carl Rogers

I hope that the ideas and suggestions in this book have helped the teacher to become more of a 'facilitator' in providing opportunities for engaging the whole class in practical activities and at the same time giving the individual pupil an opportunity to work within the scope of his or her own abilities.

The activities have offered the children an adventure in making music. If they have been enjoyably engaged in experimenting, experiencing, in decision making and in composition then they will have gained in confidence, responsibility and self-confidence. They will also have developed fundamental aural skills and an enrichment in musical awareness which will form the basis and provide the framework for many further musical activities that they might be interested in.

Appendix

SOME USEFUL MUSICAL REFERENCES

Terms and signs

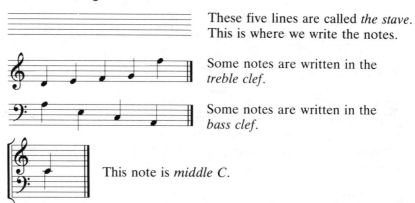

These five lines are called *the stave*. This is where we write the notes.

Some notes are written in the *treble clef*.

Some notes are written in the *bass clef*.

This note is *middle C*.

Here is middle C on the piano keyboard. This is the same note as the lowest note on the alto xylophone or metallophone.

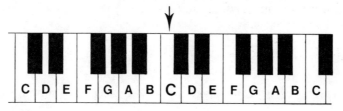

We can work out the notes, using a 'musical alphabet', writing them alternately on the lines and in the spaces. (See *Vocal Activities* for further practical explanations.)

♯ This sign is called *a sharp*.   ♭ This sign is called *a flat*.

They are used when we wish to modify a note, i.e. to 'sharpen' it (raise it a semitone), or 'flatten' it (lower it a semitone).

They are often used in various combinations at the beginning of a piece of music to indicate the *key*. This is a way of forming scales which conform to a set pattern, starting on different notes. It is an important factor in the harmony of much Western music.

Ostinato
This is a repeated rhythmic pattern:

or melodic pattern:

A rhythmic pattern or phrase is a musical shape represented only in rhythmic values:

A melodic pattern or phrase is a musical shape which uses notes of different pitches and rhythmic values:

:‖ This sign tells us to repeat a phrase or section of music.

⌢ This is a pause sign, which is self-explanatory.

Silence is represented by a sign called *a rest*:

𝄾 = ½ beat rest

𝄽 = 1 beat rest

▬ = 2 beats rest

▬ = 4 beats rest *or* 1 whole bar of silence

Instruments

Indian bells are two small but heavy brass cymbals joined by tape.

A woodblock can be either a wooden box with a slit in the side, or a hollow cylindrical-shaped piece of wood.

Gato drums come in three sizes. They are solid wooden, cube-shaped instruments. The playing surface consists of slats of wood fashioned to produce notes of random pitch.

Claves are two sticks made of very hard wood.

A guiro is usually a piece of bamboo with notches cut in the side. A stick is scraped along the notches to produce a sound.

Chimebars are metal strips mounted on resonators. They come as individual bars and can be used in different combinations.

A glockenspiel is a pitched instrument with thin metal bars. It usually comes in two sizes – the soprano and the alto. The alto glockenspiel produces lower notes.

A xylophone is a pitched instrument with wooden bars.

A metallophone is a pitched instrument with thick metal bars. Both these instruments come in three sizes: soprano; alto; bass. The smaller instruments sound higher; the larger instruments sound lower.

Bass xylophone bars are wooden bars mounted on resonators. They come as individual bars and can be used in different combinations.

Bibliography

The following is a list of useful books both for further reference and for other practical ideas:

Beaters (Schott & Co.)
Children Dancing by Rosamund Shreeves (Ward Lock Educational)
Composer in the Classroom by Murray Schafer (Universal Edition)
Creative Themes by Henry Pluckrose (Macdonald Educational)
Exploring Sound by June Tillman (Stainer & Bell)
Hear and Now by John Paynter (Universal Edition)
Music and Language with Young Children by K.M. Chacksfield (Basil Blackwell)
Music Games to Make and Play by June Tillman (Macmillan)
Musical Starting Points with Young Children by Jean Gilbert (Ward Lock Educational)
Oxford Primary Music Course, Stages 1 and 2, by Jean Gilbert and Leonora Davies (Oxford University Press)
Rhinoceros in the Classroom by Murray Schafer (Universal Edition)
Sound and Silence by John Paynter (Cambridge University Press)
Sound Ideas by K.M. Chacksfield (Oxford University Press)
Sounds Fun, Books 1 and 2, by Trevor Wishart (Universal Edition)
The Dance and the Drum by John and Elizabeth Paynter (Universal Edition)
The Music Club Book by Albert Chatterley (Stainer & Bell)
When Words Sing by Murray Schafer (Universal Edition)

Index